The Botanic Age

DEAN FALK

The Botanic Age
Planting the Seeds of Human Evolution

ÆVO UTP

Aevo UTP
An imprint of University of Toronto Press
Toronto Buffalo London
utorontopress.com

© Dean Falk 2025

ISBN 978-1-4875-4664-9 (cloth) ISBN 978-1-4875-4774-5 (EPUB)
ISBN 978-1-4875-4711-0 (PDF)

All rights reserved. No part of this publication may be reproduced, stored in or introduced into a retrieval system, or transmitted in any form or by any means (electronic, mechanical, photocopying, recording, or otherwise) without the prior written permission of both the copyright owner and the above publisher of this book.

Library and Archives Canada Cataloguing in Publication

Title: The botanic age : planting the seeds of human evolution / Dean Falk.
Names: Falk, Dean, author
Description: Includes bibliographical references and index.
Identifiers: Canadiana (print) 20240443942 | Canadiana (ebook) 20240443985 | ISBN 9781487546649 (cloth) | ISBN 9781487547745 (EPUB) | ISBN 9781487547110 (PDF)
Subjects: LCSH: Human evolution. | LCSH: Hominids. | LCSH: Baskets. | LCSH: Nests.
Classification: LCC GN281 .F35 2024 | DDC 599.93/8 - dc23

Cover design: Heng Wee Tan
Cover image: iStock.com/borchee

We wish to acknowledge the land on which the University of Toronto Press operates. This land is the traditional territory of the Wendat, the Anishnaabeg, the Haudenosaunee, the Métis, and the Mississaugas of the Credit First Nation.

University of Toronto Press acknowledges the financial support of the Government of Canada, the Canada Council for the Arts, and the Ontario Arts Council, an agency of the Government of Ontario, for its publishing activities.

For my brother, sister, and sister-in-law:
Arthur, Dana, and Dina Davis

Contents

List of Illustrations ix
Preface xiii
Acknowledgments xvii
Timeline of Key Events xix

Introduction 3
1 Baskets in the Trees 8
2 Baskets Go to Ground 26
3 Did You Make Your Nest This Morning? 50
4 From Tree Nests to Baby Carriers 70
5 First Came Wood, Then Came Stone 87
6 Babies Fall, Language Rises 110
7 What's Hobbit Got to Do with It? 128
Conclusion 142

People behind the Book 151
Notes 191
References 217
Index 245

Illustrations

1.1 Evolutionary tree of the great apes and humans   9
1.2 Chimpanzee in a "basket" in a tree   16
1.3 Chimpanzee piling up boxes to reach for fruit   23
2.1 Human versus chimpanzee hips   29
2.2 Chimpanzee big toe   31
2.3 Evo-devo walkabout   34
2.4 Dr. Ronald Clarke with the skull of Little Foot (StW 573)   37
2.5 The Botanic Age and beyond   41
2.6 Dr. Kathelijne Koops in a chimpanzee ground nest   45
3.1 Taung child killed by eagle   52
3.2 Typical sleep pattern for mature humans   54
3.3 A sleeping hut constructed by Hadza women   58
3.4 *The Nightmare* by Henry Fuseli   65
4.1 Infant monkey clinging independently to its leaping mother   72

4.2	Little chimpanzee clinging independently to its mother	73
4.3	Net baby sling from Papua New Guinea	79
4.4	Use of slings, cradles, and arms reported in the ethnographic literature	80
4.5	Woman with child in deep sitting cradle	81
4.6	Woman carrying child and other items	84
4.7	"Man the protector, woman the porter"	85
5.1	Engraved slate from Gönnersdorf, Germany	88
5.2	Commemorative stamp from Armenia	90
5.3	Reconstruction of shelters and tree canoes from Finland	91
5.4	Oldest wheel with axle from Ljubljana Marshes, Slovenia	92
5.5	Acheulean hand axe from Tanzania	94
5.6	Tip of a two-foot-long chimpanzee hunting spear	102
5.7	Wood spears from Schöningen, Germany	104
5.8	The Clacton spear, dated to about 400,000 years ago	106
6.1	The palmar grasp reflex	116
6.2	Larynx motor cortex (LMC) in the left side of a human brain	124
7.1	Paleoartist's reconstruction of LB1 ("Hobbit")	129
7.2	Possible routes of *Homo floresiensis* to Flores, Indonesia	130
7.3	Palm rafts that beached on the coast of Borneo	134
7.4	Dugout canoe from the village of Pesse in the Netherlands	137
8.1	Oscar the Bird King, created by Thomas Dambo	146
8.2	Stickwork sculpture, created by Patrick Dougherty and Sam Dougherty	147
8.3	Pueblo fiber artist Louie García at the loom	149
9.1	Cover of Rudyard Kipling's *Just So Stories*	152
9.2	Dr. Matz Larsson with his grandson	155
9.3	A pregnant woman	158

Illustrations   xi

9.4   Dr. Kathelijne Koops   161
9.5   Dr. Susanne Shultz at a wolf research center   164
9.6   Dr. Cara Wall-Scheffler   169
9.7   Dr. Helen Anderson preparing to repair a basket   173
9.8   Dr. Rebecca Biermann Gürbüz exploring an archaeological site in Turkey   178
9.9   Glenn Marshall with current drifters and completed raft   183
9.10 Completed raft for Glenn Marshall's 2020 project   184
9.11 Pueblo fiber artist Louie García   187

Preface

As you can imagine, a book that focuses on the first 3.5 million years of hominin evolution – identified here as the Botanic Age – contains a good deal of speculation. After all, this long stretch of deep time lacks an archaeological record of tools, and its fossil record of possible hominins is sparse and contentious. (Everybody wants to find a hominin – it's no wonder there are few, if any, ape fossils identified during this period!)

How, then, does one begin to explore hominin evolution from the time our ancestors split with those of chimpanzees around 6.5 million years ago to the beginning of the Stone Age 3 million years later?

Happily, there are ways. Paleoanthropologists are trained to compare the anatomies, genes, and behaviors of living humans, monkeys, and apes (especially chimpanzees, who are our closest nonhuman cousins). They also examine how individuals in different species develop as fetuses, newborns, infants, and so on. This is known as evolutionary developmental biology, or evo-devo for short.

It is also important to analyze the fossil record of undisputed hominins, beginning a little more than four million years ago. As we will see, the fossilized remains of our ancestors help us address certain questions: What do hominins' bones tell us about how they moved at different points during prehistory? Which hominin species were associated with inventing specific kinds of tools, and does the answer have implications for the evolution of advanced cognition and whether one sex played a bigger role in it?

Some of the researchers you will meet in the following pages use experimental approaches – figuring out how to make and use the kinds of tools that hominins manufactured at different points in time, or constructing seacraft from natural materials and then trying to use them to get somewhere, like Australia, which wasn't inhabited by people until around 65,000 years ago. In a similar vein, experimental primatologists have reverse-engineered arboreal sleeping nests of great apes to determine how they are constructed and the cognitive skills that may be involved in doing so.

One of the most crucial methods for addressing questions about Botanic Age hominins and how they lived is based on the realization that, for over 99 percent of their 6.5 million-year-long existence, hominins lived in nomadic bands and fed themselves by gathering plant foods, foraging for small animals, scavenging carcasses, hunting, and fishing. Margaret Mead[1] and other early pioneers of cultural anthropology traveled to, lived with, and studied contemporary people who lived under similar conditions in cultures that are sometimes called "ethnographic-period societies." Such cultures are rapidly vanishing and much (but not all) of the ethnographic research considered in this book took place before 1960. Cultural anthropologists like Mead made us aware that these cultures are enormously valuable for thinking about early evolution because their living conditions and lifestyles more closely resemble those of early foraging and gathering hominins than those of people living in postindustrial societies. It is important to emphasize that the residents of ethnographic-period societies, including the dwindling handful that still exist, were and are just as modern as people in other parts of the world; they are not and should not be thought of as "living fossils." In fact, evolutionary anthropologists generally agree that WEIRD societies (i.e., Western, educated, industrialized, rich, and democratic) "are among the least representative populations one could find for generalizing about humans."[2] As you will see, the scholars who rely on information from ethnographic-period societies in their research describe them variously as "non-Westernized societies," "small-scale societies," "subsistence societies," "traditional societies," "preindustrial societies," or combinations thereof such as "small-scale subsistence societies (4S)." All of these identifications are meant to be respectful. Moreover, those of us who are curious about humanity's past should be grateful to the residents of these cultures who were and are willing to tolerate and share information about their lives with inquisitive visitors from nontraditional, Westernized, and industrialized societies like those many of us live in.

Acknowledgments

I am immeasurably grateful to my dear friend and respected colleague Adrienne Zihlman for her numerous careful readings of various drafts of *The Botanic Age* and her constructive suggestions for improving it. Adrienne was one of the first to recognize that the earliest hominin inventions likely included botanical baby slings and that females were at least as important for charting the path of hominin evolution as males – not to mention that she is a superb comparative anatomist and paleoanthropologist.

The following scholars answered questions about their research in written interviews that appear in the "People behind the Book" section at the end of the book: Helen Anderson, Louie García, Rebecca Biermann Gürbüz, Kat Koops, Matz Larsson, Glenn Marshall, Susanne Shultz, and Cara Wall-Scheffler. Their seminal research had a huge impact on the development of the ideas in the book.

Additional inspiration came from conversations and email exchanges with Myrdene Anderson, James B. Harrod, Alice Kehoe, David Lindsay, and Bill McGrew.

Illustrations, or help finding them, were kindly provided by John Barker, Dina Davis, Elysia Poon, and Joachim Richter. Arthur Davis is my terrific webmaster (and brother) and Ryals Lee Jr. took the author's photograph for the book jacket.

Florida State University generously provided a sabbatical during spring semester 2021, which I used to work on *The Botanic Age*. I'm thankful to Sheryl Grossman for helping with my sabbatical application.

It was a thrill to give my first public talk about the Botanic Age at an event hosted in November 2023 by The Leakey Foundation. Arielle Johnson is thanked for all the hard work she put into making my visit so enjoyable.

I am deeply indebted to my editor at University of Toronto Press, Carli Hansen, and to the managing editor at UTP, Leah Connor. Jenny O'Reilly did a bang-up job copyediting the manuscript. As always, Joel Yohalem is thanked for providing thoughtful input on multiple drafts during the book's nearly five years of gestation.

Last but not least, I am grateful to my wonderful agent, John Byram at Duke City Consulting.

Timeline of Key Events

7–5 million years ago (MYA) – Late Miocene Cooling; continuous forest begins to fragment

~6.5 MYA – The Botanic Age begins; chimpanzees and hominins start to diverge

4.6 MYA – Hominins increase terrestrial bipedalism; some males sleep on ground; infants fall from mothers

~3.7 MYA – Aligned (humanlike) big toes appear in two australopithecine species; botanical baby slings likely used

~3.5 MYA – The Stone Age begins

3.4 MYA – Cut marks on bones in Ethiopia

3.3 MYA – Earliest known stone tools (Lomekwi, Kenya)

2.8 MYA – Oldowan pebble tools appear in Kenya

before 2.0 MYA – First hominins leave Africa

1.8 MYA – Acheulean hand axes appear (more sophisticated)

1.5 MYA – Fire-burned tools and bones may indicate control of fire (*Homo erectus*)

Timeline of Key Events

1.5 to 1.0 MYA – *Homo erectus* fully terrestrial (i.e., everyone sleeping on ground)

before 1.0 MYA – Hominins arrive on island of Flores, Indonesia, likely by accident via "baskets in seas"

780,000 years before present (YBP) – Polished wood plank (Israel)

500,000 YBP – Shell incised with zig-zag pattern by Asian *Homo erectus*; crosshatched patterns carved on slabs in Africa (similar to weaving patterns)

476,000 YBP – Notched interlocking logs (Zambia)

400,000 YBP – Earliest known hominin-made wood spear (Clacton-on-Sea, UK)

200,000 YBP – Earliest known use of grass bedding (Border Cave, South Africa)

~65,000 YBP – Humans arrive in Australia, likely via watercraft

30,000 YBP – Flax fibers that may have been used to weave baskets, sew

~25,000 YBP – Humans arrive in the Americas, likely via foot and watercraft

15,000 YBP – Earliest recorded baby carrier

~12,000 YBP – Beginning of New Stone Age (Neolithic)

~10,000 YBP – Earliest known dugout canoe

~5,500 YBP – Writing invented

~5,000 YBP – The Stone Age ends

5,000 YBP – Earliest known wooden wheel and axle

2,000 YBP – Paper invented in China

The Botanic Age

Introduction

HOW AND WHY DID HUMANS GET to be so clever and thoughtful? This question has long fascinated naturalists and others. Many anthropologists maintain that the first glimmerings of advanced intelligence were sparked when our early ancestors (known as hominins) began modifying rocks to make tools for butchering carcasses and using as weapons. In fact, this narrative became so entrenched that scientists referred to the span of time between the earliest evidence for deliberately crafted stone tools (currently, around 3.5 million years ago) and the beginnings of metallurgy (a mere 5,000 years ago) as the Stone Age.[1] Simply put, the received wisdom is that, when it comes to what makes humans *human*, the Stone Age is "when it happened."

This idea did not originate with anthropologists, however. It has been around since at least the time of Darwin and has even crossed over into popular culture. Recall, for example, the silent opening scene from the film *2001: A Space Odyssey*,[2] which beautifully symbolized the narrative that early hominins "received" wisdom from stone tools when a troop of early hominins communally laid their hands on a mysterious stone monolith.[3] Their newfound intelligence enabled the males to use all kinds of tools not only to prevail in a territorial fight with another troop, but also to hunt for meat. Near the end, a dominant male hurled a bone into the air that transformed, via a three-million-year leap into the future, into an orbiting space vehicle. In keeping with the title of the scene, "The Dawn of Man," the females and two chimpanzees that were reportedly used as stand-ins for hominin infants were very much in the background. This, of course, reinforced the textbook narrative of how *man* (i.e., *Homo sapiens*) became wise. Despite its male bias, though, the sequence is visually stunning, and the hominins are so accurately portrayed that it is hard to believe the film was made more than half a century ago.

Awe-inspiring as *2001* is, the film shares a major flaw with the mainstream view of cognitive evolution. Both anchor the "dawn" of advanced cognizance at the beginning of the Stone Age, now thought to have been around 3.5 million years ago. However, the hominin lineage that gave rise to humans split from the one that led to today's chimpanzees (our closest nonhuman relatives) approximately three million years before that. Although this fact is solidly supported by comparative genetics and widely accepted, the first half of the period of hominin evolution continues to be more or less ignored by scientists in the rocks-kicked-it-all-off school of thought. In fairness, there is not much in the way of relevant archaeological materials or hominin fossils from this early period. But this does not mean that our predecessors began to develop "wisdom" only after the Stone Age began.

In fact, our hominin ancestors underwent dramatic evolutionary changes during their earliest years of existence, long before any of them ever battered a rock into a useful tool. Even with a less than satisfactory fossil and archaeological record, we know that the long stretch of time that preceded the Stone Age was, literally, when our ancestors experienced their "formative years" (chapter 1). While our closest nonhuman cousins are chimpanzees, we are also close to the other two great apes: gorillas and orangutans. Of the hundreds of species of higher primates, only the three great apes construct sleeping nests in trees every day. Handcrafted sleeping nests were an important invention of ancestral apes. From what we know about our evolutionary relationship with them, we can be certain that early hominins inherited a special ability for bending and weaving branches into basketlike nests that were sturdy enough to keep them high, dry, and secure while they slumbered in the treetops. The ability to make these stationary tools (which is what they were) was a complicated business that depended on learning and, to some extent, an intuitive understanding of how the physical world works (sometimes called folk physics in humans). During the day, hominins began spending less time in the trees and more of their waking hours walking upright on the ground (chapter 2). As we will see, this profoundly changed not only their bodies but also their sleep patterns and nervous systems (chapter 3).

Many anthropologists regard the evolution of bipedalism as *the* main event that shaped hominin origins. Rather than a walk in the park, however, the journey from four-footed to two-legged locomotion was fraught with danger, as fewer babies developed the ability to cling with all fours to the underbellies of their traveling mothers,[4] the way that ape (and monkey) infants do. Without some kind of intervention, it is highly likely that an increasing number of infants would have suffered lethal falls. This was a potentially serious survival problem during the slow emergence of bipedalism, which is why some visionary anthropologists long ago reasoned that baby slings were among the very first tools ever invented[5] (chapter 4). Indeed, it wouldn't have taken much of a lightbulb moment for prehistoric mothers to invent the first baby slings, because the intuitive physics that prompted their ancestors to begin weaving night nests was already part of their nature.

As hominins continued to evolve during the Stone Age, they encountered other challenges that required them to draw and expand upon their grasp of folk physics and manipulative skills to invent new kinds of tools – not just from stone, but from the botanical materials with which they were already familiar, such as grasses, vines, leaves, and wood (chapter 5). After all, these were the materials that the earliest hominins had used to construct new inventions. Furthermore, contrary to many anthropology textbooks and at least one award-winning movie, when one takes a hard look at the scientific evidence, it becomes clear that females must have done their fair share of the inventing (especially when it came to the most game-changing innovation of them all: language) (chapter 6).

The weaving of nests and the using of sticks marked a prolonged and previously unnamed period of time that should be rightfully recognized as the Botanic Age, preceding the emergence of the Stone Age. Although "Basket Weaving 101" has often been used as a slang term for university courses that are easy to pass and teach little of value,[6] nothing could be further from the truth when it comes to hominin evolution. As we will see, the weaving of stationary arboreal sleeping nests (baskets in trees) led to the first complex portable tools invented by our bipedal ancestors (including life-saving baby slings) and to subsequent botanical inventions, including sophisticated means of transportation and other kinds of fiber-based innovations. From static hominin sleeping trees in Africa to beached trees on tropical Indonesian islands millions of years later (chapter 7), to the first deliberately built watercraft, the impact of the Botanic Age on all of hominin evolution was, in a word, unparalleled – even by the chipping of rocks into tools, which gave the Stone Age its name and its false reputation for birthing our ancestors' earliest intellectual awakenings.

The story that will unfold in the following pages draws significantly from the inspiration and discoveries of numerous scholars. Although their fields are extremely diverse, they share an intense curiosity about humanity, as well as an interest in human ancestry – be it in the deep past or more recently. For readers who would like to learn more about the science and scholarship that substantiate some of the assertions in *The Botanic Age*, a section called "People behind the Book" can be found at the end of the book, along with references and notes. In it, eight of the people whose ideas were seminal for this book answer questions that highlight their contributions. Their answers will give you a glimpse not just of the scholars' intellectual vibrancy, but also their enthusiasm.

You may recall playing rock-paper-scissors as a child. In it, two players tap their fist onto the palm of their other hand and, on the count of three, simultaneously throw it into one of three shapes: a closed fist (rock), an open hand (paper), or a V shape with the index and middle fingers (scissors). If the players throw different shapes, one wins. (If not, they try again.) Paper covers rock, rock smashes scissors, and scissors cut paper. Paper traditionally comes from wood pulp or other plant fibers. Like the exceptional scholars you will meet in this book, rock-paper-scissors players believe that plant materials beat rocks. The following chapters make the case that millions of years of basket weaving and woodworking not only preceded the Stone Age, but also played an important part in the evolutionary development of our species, laying the neurological groundwork for our later creative and technological inventions.

Paper beats rock!

1
Baskets in the Trees

AS EVERYONE KNOWS, OUR SPECIES EVOLVED with the cognitive abilities to invent amazing tools like computers and spaceships. But when and why did human intelligence originate and how did it get to this point? To begin to approach these questions we need to start with clues about our very earliest human predecessors, and that means we must begin in the trees.

When it comes to thinking about human origins, scientists have firmly established that people are descended from arboreal (tree-living) primates that lived an extraordinarily long time ago.[1] Of the many hundreds of living primates, including monkeys, the physical and genetic makeup of people most closely resembles those of the living great apes – orangutans, gorillas, and chimpanzees.[2] Great apes are, as you might guess from their name, much larger than other nonhuman primates – a very important point. The African gorillas and chimpanzees are more closely related to humans than are the Asian orangutans. Narrowing it further, geneticists have determined

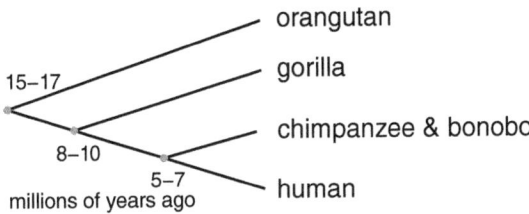

Figure 1.1. An evolutionary tree of the great apes and humans, with commonly inferred branching dates. Adapted from Groves 2018.[3]

that chimpanzees (including bonobos) are nearest to us of all. It is estimated that the human lineage of upright walking ancestors split from that of chimpanzees between five and seven million years ago (see figure 1.1).[4] This means chimpanzees are our closest (i.e., "first") nonhuman cousins, while gorillas and orangutans are more "removed" cousins.

Many of the events considered in this book took place in Africa during the first three million years after the split between chimpanzees and hominins.[5] The reader may wonder how hominin evolution can be studied prior to four million years ago, a time when there is zero archaeological record of material culture and only a handful of fossils that have been tentatively, if controversially, identified as human predecessors. Scientifically sound inferences can be made, however, by comparing the bodies, behaviors, and cognitive abilities of living great apes and humans and by studying the anatomy of potential early hominin fossils. Here is an example: Unlike the hundreds of species of monkeys, apes and humans do not have tails. Although there is a large gap in the early part of the hominin record, all available fossil hominins also lack tails. We can safely assume, then, that all hominins on the human branch of the evolutionary tree, including the earliest ones, did not have tails.

Another example: Each evening great apes make nests in trees, in which they sleep more or less horizontally, but no monkey does this. Although people rarely, if ever, sleep in trees, they retain the habit of slumbering while lying down in beds. A notable difference is that, while humans may "make" their beds, roll up their sleeping mats, or hitch up their hammocks, they do not build a new bedframe from scratch in a different location each day. Because all the great apes do, it is logical to assume that the arboreal apelike forerunners of hominins made and slept in night nests, and that their descendants brought the habit with them when they shifted to sleeping on the ground. This is when hominins began to evolve walking upright in place of walking on all fours, like great apes usually do when they happen to be on the ground. As shown in the next chapter, information about the physical evolution of hominins (from their fossils) helps pinpoint when in time this shift likely happened.

The ability to make and use tools is crucial for thinking about cognitive evolution in early hominins. But how does one do this when there is little or no record of tools until several million years after the hominin-chimpanzee split? Again, we can look at evidence gleaned from comparing tool-related behaviors in apes and humans. Many animals in nature use tools, including some insects, crabs, fish, birds, and monkeys. But wild chimpanzees use tools that are more varied than any other animal except humans.[6] These include rocks to crack nuts, twigs and branches to extract honey from honeycombs and termites from their mounds, leaves used as sponges and wipers, and natural objects that are thrown or brandished in emotional displays. As Jane Goodall famously discovered, chimpanzees are also known to prepare termite "fishing" poles well in advance – thereby demonstrating foresight once thought to be reserved for human tool makers.[7]

It is becoming apparent from the emerging discipline of primate archaeology, especially that of perishable materials,[8] that chimpanzees have community-specific techniques for termite fishing. These appear to be acquired, not just because of environmental differences, but also through social learning – reminiscent of the different ways of obtaining and processing "ethnic food" in human societies.[9] Orangutans and gorillas also use tools in the wild, although not as often as chimpanzees. Part of the reason for this may be that orangutans do not need tools to acquire bark for food, which they do when fruit is not abundant, while gorillas can simply use their great strength to tear bark off trees, rip logs apart, and remove the tops of ant nests to find food. Gorillas in zoos and parks, however, use objects left in their enclosures as tools, and as "social tools" to manipulate other gorillas.[10]

Because the great apes experienced long and separate evolutions, it is not surprising that they developed distinctive temperaments. The unique nature of each ape is enchantingly captured by Benjamin B. Beck of the Smithsonian Conservation Biology Institute:

> There is an anecdote that circulates among zoo folk describing what happens when a screwdriver is inadvertently left in the enclosure of an adult gorilla, chimpanzee, bonobo, or orangutan. The gorilla would not discover the screwdriver for an hour until accidentally stepping on it. Shrinking in fear, the ape would approach the tool only after a considerable interval, venturing a cautious, tentative touch with the back of the hand. Finding the screwdriver harmless, the gorilla would smell it and try to eat it. Upon discovering that the screwdriver was inedible, the gorilla would discard it indefinitely. The chimpanzee would notice the tool at once, seize it immediately, and try it out as a club, a spear, a lever, a hammer, a probe, a missile, a toothpick, and

everything else, except as a screwdriver. The tool would be guarded jealously, taken by dominant group members, manipulated incessantly, and discarded from boredom after several days. The bonobo would also notice the tool immediately and begin vocalizing with great intensity. The tool would be quickly picked up, inspected, and then passed from bonobo to bonobo in a flurry of excitement. The good fortune of finding such a novel item would result in copulations, genital rubbing, and an enthusiastic chorus of screams. In the midst of all of the erotic intensity, the screwdriver would be dropped and forgotten. Like the chimpanzee and the bonobo, the orangutan would notice the tool at once but ignore it, lest a keeper discover the oversight. If a keeper did notice, the ape would rush to the tool and surrender it only in trade for a preferred food. If a keeper did not notice, the ape would wait until night and then proceed to use the screwdriver to pick the locks or dismantle the cage and escape. While this anecdotal comparison is scientifically inadmissible, it is a reasonably accurate portrayal of the disposition, curiosity, manipulative ability, and propensity for tool use of the four types of great apes.[11]

Many paleoanthropologists give little consideration to the fact that the earliest hominins inherited a proclivity for using tools from the ancestor they shared with chimpanzees. As noted above, the other great apes do not appear to use the variety of tools that chimpanzees do. The types of tools and the ways they are employed are not universal – even in chimpanzees, because different communities use different kinds and combinations of tools to different degrees. Thus, it seems unlikely that the use of vegetation to fish for insects provided the foundation for the eventual emergence of advanced cognition in early hominins. Instead, the starting point likely entailed a complex and frequent use of natural materials that has been retained by all living great apes.[12] Only one behavior I know of fits this description – namely, the construction of arboreal sleeping nests. Even though other animals such as birds build nests in trees, the fact that nest building is surprisingly uniform across great apes,[13] combined with the evolutionary tree in figure 1.1, suggests that this behavior emerged independently from other animals. This would have occurred in a common ancestor of all great apes after it split from the much smaller-bodied lineage of non-nest-building "small-bodied ['lesser'] apes" (i.e., gibbons).

In a way, great apes' sleeping nests may be thought of as stationary tools. In fact, others have gone so far as to speculate that "nest building is not only properly placed within the realm of tool use … it is also the original tool that led to the mental and physical ability to use the tools we see today … Thus, the nest served as the spring board for the great leap forward in hominid evolution."[14] As we will see, early hominins likely initiated their stunning toolmaking career by using their manipulative and nest-making skills to invent new kinds of botanical tools that helped them adjust to life on the ground.

How and Why Great Apes Build Sleeping Nests

Monkeys typically sleep crouched, semi-upright, or draped across branches in trees, although some of the tinier species sleep in treeholes. Old World monkeys[15] such as baboons and macaques even have thickened pads on their rears (ischial callosities) that cushion their bottoms when they sleep in sitting positions. Unlike monkeys, great apes have highly flexible shoulders, long mobile arms that can reach in any direction, and broad chests.[16] They are also more likely to fall from trees and are at greater risk of serious injury or death when this happens because they are much larger than monkeys.[17] Fortunately, because the great apes can reach in all directions to manipulate and pull together plant materials, they are adept at constructing sleeping nests that lessen the chance of falling, at least at night.[18]

During their adult lives, orangutans, gorillas, and chimpanzees make sleeping nests every evening around dusk, which they usually occupy alone, except for mothers who sleep with their unweaned infants. Because orangutans are the most arboreal and least social of the great apes, their tree nests are likely to be separated. The African great apes, on the other hand, make their nests near each other (sometimes in the same tree) at different sleeping sites each afternoon. They may return to abandoned sites to build new nests or refurbish previous ones, especially in drier habitats where nesting materials are scarce.[19] Remarkably, reuse of the same nest sites and trees appears common enough that the scars, breaks, bends, and regrowth of branches that occur from nest construction may "inadvertently alter … favorite trees in such a way as to create branch structures prefabricated for future nest construction."[20]

Individual apes are likely to build many thousands of nests during their lives. A bonobo, for example, is estimated to build around 19,000 nests during its lifetime.[21] Most of the nests are in trees, although apes may also make less substantial nests on the ground for resting during the day. Some individuals, including very large gorillas, may even make their night nests at the base of a tree rather than in the branches. As discussed in the next two chapters, studies of unusual populations of chimpanzees suggest that a shift from arboreal to terrestrial sleeping nests may have contributed in surprising ways to early hominin cognitive evolution.

An infant great ape sleeps in its mother's nest for several years before it is weaned and learns to build its own nest by watching her.

Goodall observed an eight-month-old chimpanzee playfully attempting to make a nest on the ground, as well as one-year-olds that made rudimentary nests in trees. One-year-old gorillas and orangutans also play/practice construction or building.[22] This does not mean that nest building lacks an innate component. Human-reared and isolation-reared apes that have never seen other apes making nests have also been observed going through the motions, albeit awkwardly.[23] For example, a human-reared one-year-old gorilla named Goma "pulled down branches one by one, stood on them, and patted them down under and around her, similar to young apes in the wild."[24]

Thanks to intrepid primatologists who climb up trees to examine, measure, and disassemble tree nests,[25] a good deal is known about the general architecture and mechanics of nest building, which are similar in all the great apes. Early hominins would have built arboreal sleeping nests in the same way. Imagine one of our early ancestors climbing into a tree in the late afternoon and selecting a sturdy horizontal limb as a foundation over which she begins to pull down nearby branches. She forms a rim for the platform by bending and interweaving these branches and breaks off others that are close by to weave from the outer to inner surface of the emerging bowl-shaped nest. Perhaps she climbs in to test how it feels, and then detaches some nearby leafy twigs to line the center.[26] Many of the great apes' arboreal sleeping nests tend to look something like big leafy baskets (figure 1.2) – hence, the title of this chapter.

Despite this uniformity in nest building, there are some interesting variations in the techniques for doing so that illustrate the cognitive complexity of making arboreal sleeping nests. For example, orangutans make strategic use of the natural properties of wood by carefully selecting and partially breaking branches from the surrounding area, but not detaching them, and then bending and weaving them together

Figure 1.2. A chimpanzee in a "basket" it made in a tree. Photo courtesy of Kathelijne Koops.

to lock into the nests' foundations, which are typically constructed on stable branches or in tree forks. After that, padding is added from detached vegetative matter, and individuals may even enhance their nests with a botanical roof, pillow, or blanket.[27] A previously unknown technique for securing arboreal nests has been observed in some arboreal nests made by chimpanzees living in Uganda.[28] "Nest tying," as it is called, secures an otherwise unstable nest by looping or wrapping leafy stems, woody vines, or fronds around the tree's trunk or an adjacent tree and incorporating them into the nest's mattress.[29] In another variation from elsewhere in Africa, chimpanzees overlay the basic circular platform of their nests with branches manipulated into a triangular shape that is then covered with a mattress of overlapping branches.[30]

Night nests offer the same benefits to apes that monkeys get from sleeping in trees: camouflage from predators, protection from biting insects and creepy creatures like snakes (which apes are wary of), conservation of body heat, proximity to individuals sleeping nearby (orangutans may be an exception), and shelter from inclement weather and blowing winds – in other words, they provide a comfortable, safe place to sleep. The arboreal sleeping nests of great apes have another crucial safety feature. They prevent sleeping apes from falling to the ground, which is a significant risk as shown by numerous healed breaks in the skeletons of primates in museum collections.[31] Great apes appear to be cognizant of this risk: "Mothers clearly guard against their infant falling: when moving in the canopy they carefully support the offspring with one hand or foot, or in the groin pocket by flexing one thigh. They wait to help if older offspring struggle to cross a gap between branches, and quickly grab the youngster if it looks like [it is] falling … To avoid falling, chimpanzees must be constantly vigilant while moving around in trees; careful planning of arboreal routes might have been a selection factor in the evolution of self-awareness in large-bodied ancestral apes."[32]

The fact that apes seem to grasp that tumbling from trees can cause grave consequences suggests that chimpanzees might, at least to some degree, understand causality. But do they?

Folk Physics

In modern *Homo sapiens*, knowledge about cause and effect in the natural world is sometimes called "folk physics." Folk physics is defined as a spontaneous "understanding of the nature of physical objects and intuitive … prediction of physical events in nature. It can be seen as a naïve or common sense approach ... [which is] very different from the theoretical approach of scientific physics."[33] For example, we all use subliminal mental intuition (at least in part) when we select a container to hold a certain quantity of food or liquid, push harder to move heavier objects, widen our stride to cross a gap, or go through the motions of delivering a basketball into a hoop.[34] There is no need to understand the nitty-gritty Newtonian physics behind such acts – we just know what to do.

People are not born with such intuitions, however. They acquire them in a certain order during the first year of life.[35] By four months of age, human babies understand that objects are bounded and continuous in time and space; a month later, they can tell liquids from solids; at around six months, infants become aware that objects can strike and launch other objects; by eight months, they have an intuitive sense of the impenetrability of solid objects; and by the end of the first year, human babies have become sensitive to the approximate placement needed for an object's center of mass relative to the edge of a supporting surface to prevent it from falling. These findings suggest that very early in life people develop a mental framework for implicitly predicting how physical events will unfold – something that has been likened to an "intuitive physics engine."[36]

Chimpanzees appear to develop a similar engine for making physical inferences, as suggested by their performances on a limited battery of cognitive tests given to chimpanzees, orangutans, and 2.5-year-old human children. The chimpanzees and children had very similar cognitive skills when it came to dealing with the physical world, but the children were far ahead of the apes in tests of social cognition, which led to the conclusion that "2-year-old children's cognitive development in the physical domain [is] still basically equivalent to that of the common ancestor of humans and chimpanzees some six million years ago ... but their social cognition [is] already well down the species-specific path."[37]

But do chimpanzees' construction of sleeping nests and use of tools indicate a grasp of folk physics that goes beyond that of a human toddler? Did we as humans inherit our intuition about the physical world from apelike ancestors? Cognitive psychologists argue vigorously about these questions. In 2000, cognitive scientist Daniel Povinelli published an influential book based on years of experiments on chimpanzees, titled *Folk Physics for Apes: The Chimpanzee's Theory of How the World Works*,[38] which has inspired countless other researchers and continues to provoke discussion and controversy to this day.[39]

Povinelli and his collaborators sought to determine what chimpanzees consciously understood about how and why tools work by testing their ability to use them to retrieve food from problematic places (e.g., by manipulating sticks to obtain an item from a clear tube with a trap in it) and assessing their aptitude for selecting appropriate tools to retrieve edible goodies (e.g., by choosing a rigid rather than a flimsy rake to collect an apple). The purpose of these and numerous other experiments was to learn if chimpanzees form ideas about unobservable causal mechanisms related to tool use such as force, physical connection, weight, and shape. In the end, Povinelli and his collaborators concluded that "chimpanzees possess nothing akin to the human capacity to generate theories."[40] Nonetheless, Povinelli observed that chimpanzees could reason very intelligently about tangible objects and events that they perceived, even if they did not ruminate about the unobservable physical world. He also speculated (quite reasonably, I think) that chimpanzees do not formulate abstract concepts like gravity and force because, lacking humanlike language, they have no words with which to think about or express such concepts.

If you noticed a disconnect between Povinelli's use of the term "folk physics" in the title of his book and the definition quoted at the beginning of this section, which entails an intuitive understanding of physical events without any theorizing, you are correct. The earlier definition is more relevant here for the simple reason that chimpanzees (rather than two-year-old children) currently offer the best model for speculating about the range and variation of the cognitive abilities in the 6.5-million-year-old common ancestor we share with them. Once we determine the extent to which chimpanzees understand how the physical world works, a sense of "naïve physics," we can speculate about how that sense influenced cognitive evolution during our ancestors' long journey from an arboreal to a fully terrestrial life – a journey of several million years that likely seeded the later invention of language[41] (chapter 6). Although determining whether chimpanzees have nonverbal insights about the physical world is a tall order (and a question scholars continue to debate), much of the answer was actually worked out over a century ago. All we need to do is pull the information out of the mothballs.

Ape Physics

In 1917, the German psychologist and cofounder of Gestalt psychology, Wolfgang Köhler, published the results of his experiments on problem-solving in nine chimpanzees housed at the Anthropoid Station in the Canary Islands off the northwest coast of Africa.[42] The nine chimpanzees included one adult female, five juvenile females, and three juvenile males.[43] Most of Köhler's experiments involved chimpanzees using tools like bamboo poles and boxes to try to obtain foods that had been placed out of reach, such as a banana suspended from a ceiling. His research became widely known and an English translation of his book, *The Mentality of Apes*, appeared in 1925.

The chimpanzees were good at obtaining prized foods in all kinds of test situations, either because they did it automatically (e.g., detoured around visible barriers to reach their goal, pulled on a string to retrieve an attached goody, or used a stick to rake in a treat when the two objects could be seen simultaneously) or through trial and error (e.g., eventually moved a box to serve as a step stool underneath a suspended fruit). Once a solution like moving the box was hit upon, it eventually spread to others in the group and became a "fashion." From Köhler's many experiments, it became clear that sticks were (and are) the chimpanzees' equivalent of all-purpose Swiss Army knives. They used sticks to poke, dig, throw, play, gather ants, stab fowls, and act as spoons, levers, and weapons. Some of these uses foreshadowed later findings that chimpanzees use sticks to feed on ants and hunt lesser bush babies.[44]

Unlike modern comparative psychologists, Köhler explicitly refrained from exploring whether chimpanzees were influenced by "factors not present" or whether they were ever occupied with "things 'merely thought about.'"[45] Nevertheless, he was determined to find out "whether *any* of [the chimpanzees'] actions ... [were] ever guided by insight."[46] To Köhler, an animal behaved with insight if it approached a problem (such as how to get out-of-reach food) by getting a visual lay of the land and then proceeding to deal with it in a smooth, continuous, and seemingly spontaneous manner.[47] In the final analysis, Köhler concluded that most of the chimpanzees' solutions lacked insight, but noted fascinating exceptions.[48] For example, after a juvenile male named Sultan spontaneously invented the "jumping pole" for fun, he eventually used it as a kind of vaulting pole to grab treats that were suspended overhead. Use of this invention spread with varying degrees of success to the rest of the group. Sultan also accidentally discovered how to combine two hollow bamboo sticks into one long pole while playing, and then used it to acquire otherwise unreachable treats, leading Köhler to conclude that Sultan "makes use of the double-stick technique intelligently, and the accident seems merely to have acted as an aid ... which led at once to 'insight.'"[49]

Köhler distinguished two kinds of physics. One was the physicist's, encompassing center of gravity, moments of force, and so on. The other was the physics of "ordinary men."[50] Köhler made it clear that the latter type of physics (folk physics) was intuitive, rather than a theoretical exercise, not only in chimpanzees, but also in humans: "The chimpanzee uses a lever in exactly the same manner as a man. Of course, the apes have no *knowledge* of the relations between force, work, direction, etc., – the factors governing the physical aspect of leverage – but the [human] carrier, who lifts his broken wheel by placing a lever under it, hardly knows more about physics. There must be a kind of purely concrete and practical 'sense' of elementary implements, arising from the optical and motor functions of these primitive creatures."[51] In other words, like humans, the lever-using chimpanzees had a bit of a nose for intuitive physics.

However, Köhler also emphasized that chimpanzees were not good at balancing external objects on the ground, which had become apparent when the chimpanzees attempted to build climbing towers by stacking boxes on top of each other. Even Grande, the individual who was best at such endeavors, had frequent mishaps. She (and the other chimpanzees) seemed to understand the height and approach that were needed to reach desired objects, but not how to build her step stool so that it would not wobble when she was aboard (figure 1.3). Chimpanzees achieved the towers "at best by chance, and, as it were, by the 'struggle for steadiness.'"[52]

Figure 1.3. A chimpanzee piling up boxes to reach a suspended banana (just out of sight in the illustration). From Köhler's *The Mentality of Apes*. By Wikimedia Commons, CC BY-SA 4.0.

Köhler reasoned that, although chimpanzees had little or no ability to apply folk physics to such problems, they had "a third kind of sense – that of their own bodies." Thus, when a chimpanzee stood on an insecure tower that would strike fear into the hearts of onlookers, it would counteract "the first suspicious wobbling of the structure ... by an instantaneous altering of the balance of the body, by lifting his arms, bending his trunk, etc." Further, Köhler believed such balancing was purely physiological and, thus, did not entail insight.[53] Instead, the chimpanzees' remarkable ability to navigate wobbly towers of boxes and fling themselves brashly through trees depended on an exquisitely honed sense of balance (their own, not the boxes') and an awareness of their position in space, facilitated by the vestibular system of the inner ear.[54] So Köhler's point is well taken – while humans are better at intuiting how to place objects on the ground, chimpanzees excel at monitoring the equilibrium of their bodies while they negotiate the three-dimensional arboreal world. And why shouldn't they, given that they spend much of their time in trees (where avoiding a fall can make the difference between life and death), while we humans navigate much more securely along the ground.[55]

As we have seen, Köhler's research (often seconded by more contemporary scholars) showed that chimpanzees not only develop some intuition about how the physical world works, but also sometimes appear to use that implicit knowledge to arrive spontaneously at insightful, if atheoretical, solutions.[56] He also remarked that "so far we have not been able to tell how far back and forward stretches the time 'in which the chimpanzee lives,'"[57] a question that continues to perplex contemporary researchers.[58] It is now thought that chimpanzees employ at least some foresight when it comes to using tools. For example, chimpanzees in the Taï forest of West Africa appear to remember where to collect stones that they then transport to distant sites and use to crack open nuts.[59] Chimpanzees elsewhere also transport botanical tools from one beehive to another to obtain honey.[60] My favorite, if anecdotal, example of ape planning is a lone male chimpanzee in a zoo that repeatedly gathered and hid stones before human spectators arrived and then threw them at the visitors.[61] So, yes, chimpanzees (and likely the other great apes) show some foresight when it comes to using tools, but not nearly to the extent that humans do.[62]

It is important to remember that not all great apes fish for termites, dip for honey, or use rocks to crack open nuts; in fact, some kinds of tool use, like the bashing of nuts with rocks, are quite rare. But, having highly manipulative hands, all great apes *do* use the ape equivalent of Swiss Army knives – wooden sticks – to explore their environments and pursue various goals. And all of them weave sleeping nests from leafy branches and vines. Although weaving was likely a key thread during the unfolding of human cognitive evolution, there is no archaeological record of it until much more recently. Nonetheless, proxies for its first glimmers in hominins can be seen in the great apes. Köhler observed a female chimpanzee "'weaving' and carefully plaiting straws through the wire interstices ... [S]he thrust a strip of banana leaf through a wire mesh, laboriously drew the end back through another mesh, tied the two ends together, and continued in the same way ... I often thought that she was about to begin a deliberate, though rudimentary, constructive effort, a form of manual craftsmanship, but she could never be induced to continue."[63]

We have now laid the groundwork for bringing our earliest predecessors down from the trees. It seems likely that these pioneers took to the ground with an ability to use wooden sticks in various ways and weave sleeping nests from branches and vines, and with at least as much intuitive understanding about the physical world as Köhler documented for living chimpanzees. As we will soon see, the weaving of nests and the use of sticks and other vegetation marked the beginning of a prolonged interval that is recognized in this book as the Botanic Age, preceding the emergence of the similarly long Stone Age. As subsequent chapters will show, the weaving of arboreal sleeping nests (baskets in the trees) led not only to the first complex tools invented by our bipedal ancestors but also to subsequent fiber-based inventions.[64]

2

Baskets Go to Ground

ALTHOUGH FOSSILS ARE CRUCIAL FOR LEARNING about the earliest hominins, they are not as useful as one might think, because scientists often disagree about whether a particular specimen represents a hominin or an ape. The relevant fossils are few and far between and are usually quite fragmentary. Scientists can also be biased toward interpreting finds as hominins rather than apes, partly because hominins garner more grant support and publicity – in other words, they are "sexier." Some paleontologists may use new discoveries to confirm their previously published ideas or to cast doubt on the theories of academic rivals. Elsewhere, I have attributed this interesting, if confusing, situation to what I call paleopolitics.[1]

So how does one determine if a primate fossil is an ape or a hominin? The answer depends very much on which parts of the body happen to have become fossilized. Because teeth are the hardest parts of a skeleton, they fossilize well and are relatively common in the archaeological record. Apes and humans have different-looking teeth and jaws, which changed substantially during hominin evolution. A fossil primate with huge jutting canines ("dog teeth"), for example, would be identified as an ape rather than a hominin since hominins' canines reduced in size and became more vertical early during their evolution. Unfortunately, things are less clear for fossils representing other parts of the body because most of them are quite fragmentary. Discoveries of a relatively large part of a fossilized skeleton or even just a sizable part of a skull are rare, celebrated events.

Despite the paucity of informative fossils from the first three million years after hominins and chimpanzees split, comparisons of these rare finds with skeletons from living apes and people suggest that the single most important thing that nudged early hominins along a novel evolutionary path was the evolution of upright walking. Hominins slowly evolved a habit of walking upright on two legs when they were on the ground, a form of locomotion that scientists have dubbed "habitual bipedalism." Human bipedalism differs markedly from the four-footed knuckle-walking of our semiterrestrial (partly ground-living) first and second cousins, chimpanzees and gorillas, respectively. Human walking also differs from the four-handed clambering through trees of our most removed great ape cousin, the highly arboreal orangutan.[2] Of course, this does not mean that great apes never walk on two legs. In fact, they are perfectly capable of standing up on two legs and walking bipedally when they are on the ground. But they do not walk with the long, slow steps or in the fully upright and graceful manner of people. Indeed, humanity's fluid striding gait sets it apart, not just from the great apes, but from all the hundreds of other nonhuman primate species.

How Humans and Apes Walk

Before exploring the emergence of habitual bipedalism in early hominins, we need to understand the anatomy of walking. It is obvious that walking entails stepping forward one foot at a time with alternating feet. If you get up and stroll around other things will become apparent: As a walker stands on one leg, they bend the knee of the other leg, lift the foot, and swing it forward until the heel strikes the ground one step in front of the body. But the standing leg is not just static. In addition to supporting the body's full weight while the other leg is swinging, the standing leg helps propel the body forward by transferring the walker's weight across the bottom of the foot from the heel to the ball to the tip of the big toe. Finally, the big toe thrusts the body forward as it pushes the foot off the ground to begin the next swing forward. Technically, a walking cycle occurs from one heel strike to the next on the same side. Thus, when one leg swings, the other rolls the body's weight forward across the bottom of the foot.[3] Essentially the walker is propelling their center of gravity forward with each step and catching up with it by moving the alternate leg forward. This balancing act is the reason one scientist long ago famously quipped that "human walking is a unique activity during which the body, step by step, teeters on the edge of catastrophe."[4]

We can also think of walking as the result of muscles acting on bones. Take the standing leg, for example. When you swing one leg forward, your standing leg becomes stabilized so that you do not fall over. This happens because two gluteal muscles that run from the pelvis, over the hip socket, to the top of the femur (thigh bone) – the gluteus medius and gluteus minimus – contract when the body is supported by just that leg (while the other one is swinging forward). This keeps the hip joint from buckling. If you put your hands over

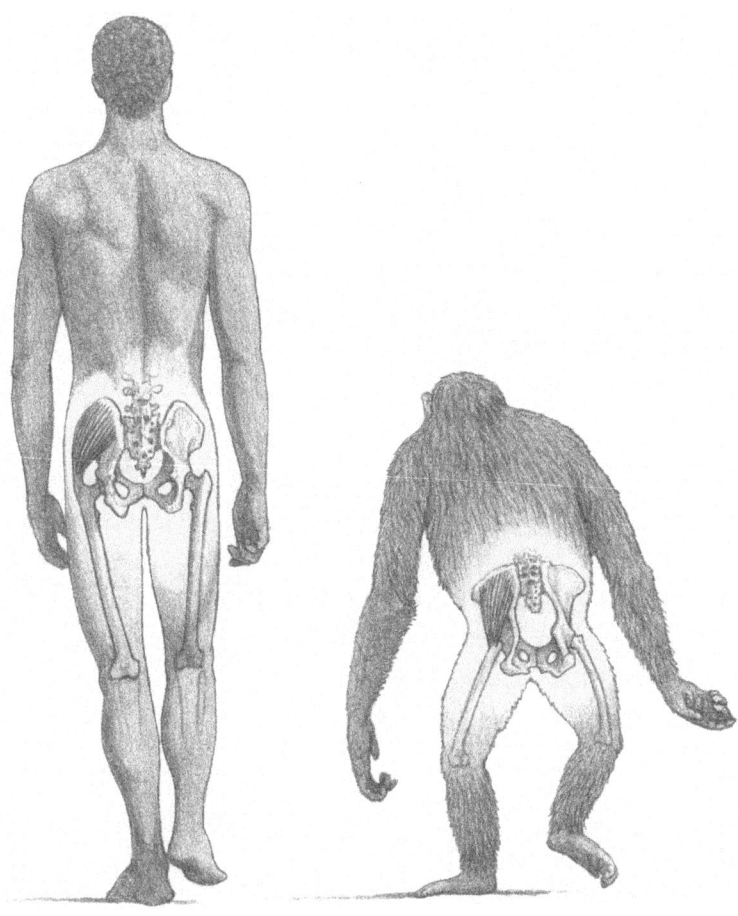

Figure 2.1. Humans do not rotate their hips as much as chimpanzees when they take steps because two of the three gluteal muscles tighten over the hip socket of the standing leg, thus stabilizing it while the person's weight is being shifted forward. Illustration by Mauricio Anton, 1998.

your hip sockets and take a slow walk, you should be able to feel the muscles on each side alternately tighten and relax. Humans have a shortened bowl-shaped pelvis, which places these two gluteal muscles (unlike the third, gluteus maximus) more toward the sides in people than apes (figure 2.1). Because these muscles are more posterior in chimpanzees and gorillas, they extend their legs toward the back when they walk instead of locking their hips. In short, the reason why apes look awkward to humans when they walk upright is because their hip muscles are not positioned to clamp down over their hip sockets.

It is not just hip anatomy that enables walking in humans. Comparisons of the skeletons, muscles, and movements of apes and humans show that the size and shape of many parts of the human skeleton are unique in ways that support a striding gait.[5] For example, human feet have high arches that transfer the body's weight forward across the bottom of the foot when walking, and big toes with expanded tips that help push the feet off the ground when they begin their swings. Unlike apes whose grasping big toes stick out to the side (figure 2.2), the relatively large but inept big toes of humans are lined up right next to the others. Put another way, human feet have become unskilled weight-bearing appendages compared to the more handlike feet of apes.

Other skeletal features are also associated with humans' ability to remain vertically balanced over an ever-changing center of gravity, countering the tendency to teeter "on the edge of catastrophe." The angle of the thigh bones places the feet and knees under the center of gravity below the pelvis, and the slightly s-shaped curve of the backbone helps a walker keep their balance. So does a head that is centrally perched on top of the spine rather than attached to it near the back of the skull like those of apes. Importantly, humans also lack certain features that reflect the arboreal lifestyle of apes, such

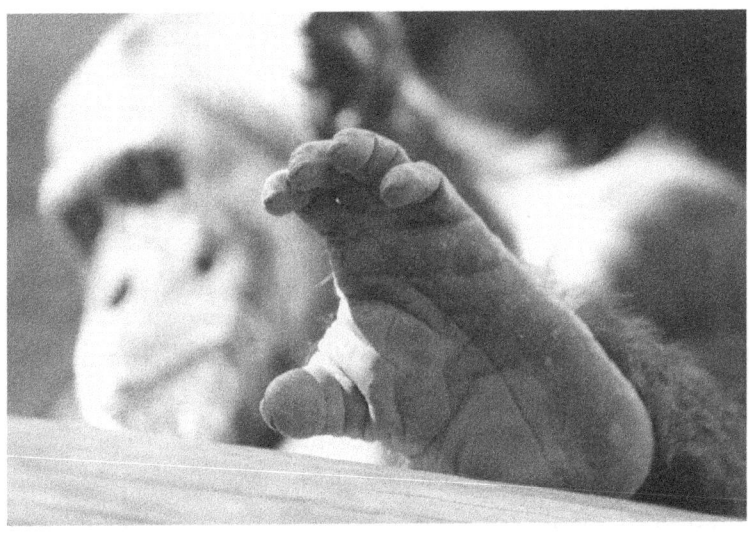

Figure 2.2. The big toes of great apes like this chimpanzee stick out away from the other toes, unlike human big toes. By AZAdam, CC BY-SA 2.0.

as extremely long arms and upturned shoulder sockets that facilitate hanging by the arms and moving through the trees. What all of this means is that signs of bipedalism are stamped from head to toe throughout the human skeleton.

That's how our skeletons are now, but how did they get that way? Clearly, it didn't happen in one fell swoop. It's clear that human skeletons were modified by natural selection in response to the emergence of bipedalism over time. But over how much time and in what order did our skeletons become modified? Strange though it may seem, we can begin to explore this question by comparing the development of infant apes and humans, an important method used in the field of evolutionary developmental biology (nicknamed evo-devo).[6] Watching babies become bipedal is not only great fun (which you can confirm by watching videos posted online by proud parents), it also provides clues for untangling how walking on two limbs rather than four may have emerged during hominin evolution.

Baby Walkers

Although certain stages in the motor development of children are generally regarded as universal, there is, in fact, a tremendous amount of variation in the ages at which milestones appear. This is due as much to infants' home environments and cultures as to their genes.[7] Stages that we think of as typical may appear out of order or not at all. Crawling, for example, is rare in cultures that refrain from putting infants on the ground for reasons of hygiene or safety. And babies may be slower to reach certain milestones in societies that keep them tightly wrapped in cradles, as is typical for some cultures in Central Asia.[8] Psychologist Karen Adolph observes that "most Western caregivers hold newborns like a fragile carton of eggs"[9] in contrast to infants in some Caribbean and African cultures where babies are deliberately exercised by being lifted by their arms or ankles, tossed, and swung around. Infants in these cultures tend to walk earlier than Western or Asian ones. In other words, the milestones for motor development in human babies aren't as universal as many parents and medical professionals may think.

Discussions about how humans learn to walk often emphasize that one foot is placed in front of the other, which can give the impression that walking is a straightforward linear activity and that it evolved as such. However, nothing is further from the truth. If you've watched a toddler trying to walk, you may have noticed that it is an erratic process. Newly bipedal infants take stuttering steps, fall, pick themselves up, lurch here and there, and stumble in various directions, including backward. (They're not called toddlers for nothing.) Nor do new walkers necessarily move in a straight line or aim for a particular destination, although walking toward someone's outstretched arms is common. As Adolph and her colleagues discuss, infants' gaits become faster, longer, and their steps narrower as they mature. "Because infants' bodies and environments are continually changing, relying on simple alternating leg movements is not viable. Instead, walking is a creative act."[10] The emergence of smooth and coordinated walking in children depends on maturation of neurological connections within the brain and the associated development of motor reflexes in the body. As with many skills like dancing or riding a bike, learning to walk is a dynamic process that benefits from practice.

Despite wide variation in the timing of children's development and the happenstance way in which they learn to walk, comparing the maturation of their motor skills with that of apes lays the groundwork for interpreting how fossil hominins moved. Although relevant research on apes is limited,[11] there is enough of it to show that chimpanzees develop approximately the same milestones as our babies, and in the same order. Thus, apes and people sit before they stand (on two or four legs for chimps, two for humans) and stand before they walk (again, on four legs for chimps, two for humans). According to one summary, chimpanzee infants sit independently, stand, and walk at a younger age than human babies do[12] (figure 2.3). These findings reflect the important fact that human babies take considerably longer to mature than their closest nonhuman cousins. Our babies have evolved to be late bloomers, and as a result they are extremely dependent on caregivers long after ape infants would be. The slowing of physical development in our ancestors' babies had

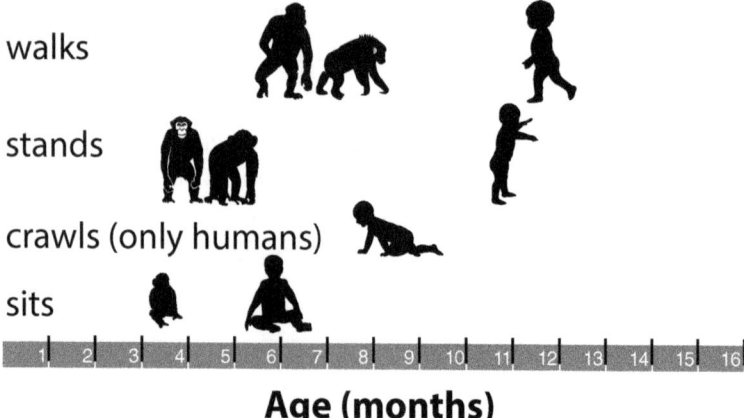

Figure 2.3. Evo-devo walkabout! Average ages by which human and chimpanzee infants sit, stand, and walk without support during their physical development.[13] Chimpanzee infants develop these milestones earlier than humans. Unlike apes, many humans experience an added crawling stage. Illustration by Dean Falk.

profound effects on the trajectory of human evolution. For now, suffice to say that the delayed physical development of human babies is one of several hand-me-downs associated with the emergence and refinement of bipedal walking millions of years ago.

It has long been known that human babies need to develop leg strength and balance before they can walk independently. "Simply put, infants cannot walk before they are able to maintain body weight and keep balance on one leg while the other leg swings forward."[14] Although chimpanzee and human infants eventually sit, stand, and walk, not all humans learn to crawl, and chimpanzees (and other apes) never do. Crawling appears to have emerged during hominin evolution, and one of its likely benefits is that it strengthens the legs in preparation for bipedal walking. Quadrupedal chimpanzees do not have this stage, probably because they strengthen their little legs by standing on all fours while keeping their torsos relatively upright and their eyes facing forward (see figure 2.3), which a human baby cannot do because its legs are too long. Little humans thus acquire an ability to support their weight (and eventually crawl) on their hands and knees because they must fold their long legs under their bodies in order to keep their torsos and faces oriented higher than their back ends, all the better to see where they are going. This added crawling stage stretches the delay between sitting and standing in humans compared to apes. Although paleoanthropologists do not pay much attention to crawling per se, its emergence in humans suggests that changes in the growth and development of legs was likely one of the key factors in the prolongation of human physical development during hominin evolution.[15]

What the Fossil Record Shows about the Rise of Bipedalism

Paleoanthropologists use their knowledge of ape and human anatomy to examine fossils for clues about when and how humanlike bipedalism arose. Because some scientists are biased toward identifying fossils as hominins rather than apes, we need to be very clear about what, exactly, qualifies a biped as "habitual." We know that a key feature associated with a striding walk is a big toe that is aligned with the other toes. Although apes are capable of awkward bipedal walking when they are on the ground, the wide angle at which their big toes project from the rest of the foot (see figure 2.2) is a dead giveaway that they do not walk on two feet as humans do – in other words, they are likely to be habitual quadrupeds, not bipeds. This must be kept in mind when seeking the earliest roots of humanlike walking.

This brings us to a fossil nicknamed Little Foot. This relatively complete skeleton of an adult female represents an early species of australopithecine (*Australopithecus prometheus*) that lived around 3.67 million years ago. What makes this fossil special is that her left foot, a good portion of which was recovered, had a big toe that was not divergent, making it the oldest known foot fossil with a big toe aligned in a humanlike manner.[16] Little Foot had accidentally fallen into a natural death trap through a shaft that led to the stark interior of a cave (called Silberberg Grotto) far below the earth's surface in South Africa. Curiously, bones of a very large monkey were found underneath her skeleton. This prompted Little Foot's discoverer, paleoanthropologist Ronald Clarke (figure 2.4), to speculate that, right before she fell, she might have been competing with a large prehistoric baboon for figs in a tree that overhung the shaft, causing them both to tumble to their deaths.[17] Similar conflicts have been reported for living chimpanzees and baboons, and trees that overhang shaft entrances are well-known sources of remains found in deep caves.

Clarke discovered Little Foot through a serendipitous sequence of events that began in 1994 when he found a forgotten box of fossils stored in a laboratory. Inside the box were a number of bones from one australopithecine's left foot. Over two years later, he came across more foot bones and a small broken-off end of a right lower leg bone in another storage box at another lab, which he realized were from the same individual. He then returned to the boxes at the first laboratory where he found still another leg bone from the same individual. (Needless to say, Clarke is gifted with an amazing visual memory.) These foot and leg fossils had been recovered from rubble blasted from Silberberg Grotto by lime miners in the 1930s, so

Figure 2.4. Dr. Ronald Clarke with the skull of Little Foot (StW 573) still partially embedded deep within a cave at Sterkfontein, South Africa. Photograph taken by Dean Falk in 2008.

Clarke reasoned the rest of the skeleton was probably still entombed in its rocky walls. To find out, his team searched the interior surface of the cave for the end of an embedded bone that would match one of the fragmentary leg bones he'd discovered. Astonishingly, they found an exact match after only two days of searching in 1997. At that point, painstaking exposure, excavation, cleaning, and reconstruction of the skeleton began. The block of rocky sediment that contained Little Foot's skull was finally lifted to the surface in 2010.

Little Foot was a short adult australopithecine, around 4'3". She had relatively long legs in proportion to her arms compared to apes, but not to the same extent as humans. However, other features of her skeleton were decidedly apelike, such as longish arms, upward turned shoulder sockets, and the anatomy of the upper arm and shoulder girdle.[18] Together, Little Foot's features suggest she possessed "adaptations to arboreal behaviors, especially those with the hand positioned above the head."[19] All the better to climb after figs! But make no mistake, despite spending time in trees (and probably sleeping there), Little Foot's little foot shows that by 3.67 million years ago at least one species of australopithecine may have been evolving into a habitual biped. Although this doesn't mean that *Australopithecus prometheus* was a direct ancestor of humans, Little Foot provides an oldest known date (so far) for the emergence of an aligned big toe that could have led to further evolution of the graceful striding gait that we see in people today.

Little Foot wasn't the only australopithecine to sport a relatively progressive big toe. Another hominin (*Australopithecus afarensis*) that lived far to the north of *Australopithecus prometheus* in East Africa 3 to 3.7 million years ago also had aligned big toes (although perhaps with more mobility than those of humans) as well as some other humanlike features of the foot. However, other toes of this australopithecine were long and curved like those of apes, which, according to foot expert Jeremy DeSilva and his colleagues, was consistent with feeding and sleeping in trees, despite the strikingly humanlike foot.[20] In keeping with this, *Australopithecus afarensis* shared Little Foot's somewhat elongated leg compared to the length of the arm as well as apelike features of the upper arm and shoulder.[21]

The most famous *Australopithecus afarensis* fossil is the relatively complete 3.2-million-year-old skeleton of a 4.5-foot-tall adult female nicknamed Lucy. Lucy hailed from Ethiopia and, like other fossils from her species, shared some features with the earlier-living Little Foot that indicate bipedalism was an important way of moving on the ground for both species.[22] As noted, both species were also characterized by upper body anatomy that facilitated moving, feeding, and resting in trees. At the time she lived, the site in Ethiopia where Lucy was found would have been a grassy woodland with many big trees. Much like Little Foot's fatal fall, paleoanthropologist John Kappelman believes severely broken bones in Lucy's skeleton (including greenstick fractures) indicate she died because she fell from a tall tree to a hard surface: "[The] impact progressed from the feet and legs to the hip, arms, thorax, and head ... When examining fossil taxa, such as *Australopithecus afarensis*, that appear to have practiced both terrestrial and arboreal locomotion, we suggest that the adaptations that facilitated bipedal terrestrial locomotion compromised the ability of individuals to climb safely and efficiently in the trees; this combination of features may have predisposed these taxa to more frequent falls from height."[23]

It is plausible that living humans who are wary of heights and falling (acrophobia) inherited this condition, which is known to run in families, because it helped keep their predecessors alive – that is, the caution inspired by such a fear was, and is, adaptive. We will return to the gravitational perils associated with the evolution of habitual bipedalism later when we discuss an important fossil of a baby australopithecine from Ethiopia.

Meanwhile, paleoanthropologists who have meticulously studied the hundreds of individual foot bones of purported hominins, in addition to the few relatively complete feet scattered across the fossil record, conclude that "burgeoning variation ... is indeed evidence for hominin diversity and experimentation in bipedalism."[24] Apparently, although all the species that have been identified as possible early predecessors of humans seem to have been adept at moving through trees, they walked bipedally in different ways when they were on the ground. In fact, many fossils (even a few with aligned but relatively flexible big toes) had overall foot anatomy that suggested their bipedal gaits were a far cry from those of living humans. Compared to humans, some species seem to have walked bipedally with comparatively floppy feet, others may have walked in manners that favored the outside edges of the feet, and some appear to have had gaits that were rather flat-footed.[25]

To summarize: What these two skeletons show is that by the beginning of the Stone Age (currently estimated at around 3.5 million years ago), at least two australopithecine species from different ends of Africa may have been well on their way to evolving a striding bipedal gait, even though they continued to spend time feeding and sleeping in trees. Although it is entirely possible that, despite their beautiful big toes, neither species was directly ancestral to humans (a highly contentious matter), the fossil record shows that the evolution of habitual bipedalism was a multi-million-year affair that repurposed the body in a sequential manner beginning with various aspects of the feet (not just modified big toes), progressing up to include the hips, and eventually manifesting in the arms, shoulders, and carriage of the head. The feet went first; the rest of the body followed, becoming fully modified for humanlike bipedalism much later when hominins shifted to full-time life on the ground (more on this in chapter 4). By then, our predecessors were making stone tools and had, thus, entered the Stone Age. But it's the three million or so years *before* the start of the Stone Age that interest us here.

The Botanic Age

The murky gap between about 6.5 and 3.5 million years ago is identified in this book as the Botanic Age[26] (figure 2.5) because at that

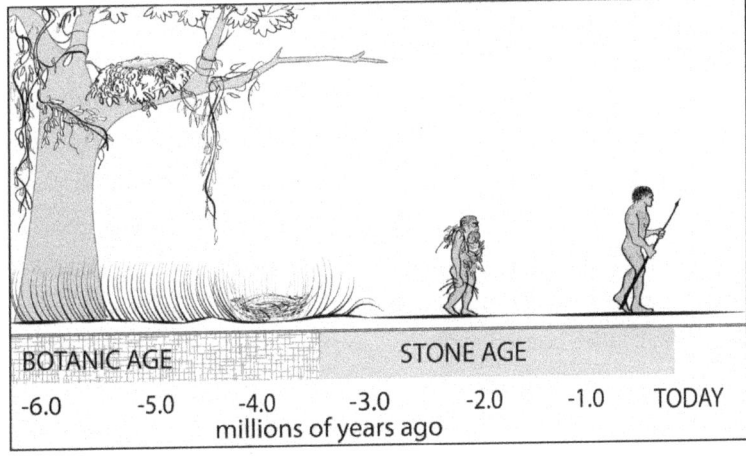

Figure 2.5. The Botanic Age and beyond. The Botanic Age lasted from the time hominins originated (~6.5 million years ago)[27] until the start of the Stone Age (~3.5 million years ago). Illustration by Dina Davis.

time hominins were still dependent on plant materials to make sleeping nests and other tools, similar to the living great apes. Together, the comparisons of modern human adults and babies, great apes, and fossilized bones of early hominins show that the most definitive feature of hominins – bipedalism – arose during the Botanic Age and was still being fine-tuned during the early part of the Stone Age. What is missing, however, is an understanding of *why* bipedalism emerged in the first place. Before we can address this, we need to consider how environmental conditions changed during the early part of the Botanic Age and think about how these changes impacted the habitats and behaviors of hominins.

For starters, it is widely accepted that the common ancestor of chimpanzees and humans inhabited heavily wooded African forests that began to decrease and break up in response to climatic fluctuations during the Late Miocene Cooling between seven and five million years ago,[28] just as hominins were in the process of diverging from their apelike ancestors. Forest restructuring continued after this initial period of cooling so that what had once been a more or less continuous equatorial forest eventually fragmented into many different forest types that still exist today in Africa (dry East African, wet central African, coastal East African, and so on).[29] Rather than being a straightforward process, however, climate change during the Plio-Pleistocene oscillated between warm wet conditions, during which African forests expanded, and colder drier conditions, when forests contracted.[30] After one cycle of warming, a renewed period of cooling occurred around 3.6 to 1.4 million years ago as the Pliocene transitioned into the Ice Age (Pleistocene), and Africa saw an increase in more open "mosaic" habitats. Thus, "at the scale of the African continent, the Late Miocene Cooling is thought to have triggered a progressive aridification, and overall the Late Miocene palaeovegetation records depict a trend to more dry open habitats and the rise of the grasslands."[31] It was in these mixed habitats where many, if not all, hominins eventually polished their bipedal gaits and likely shifted to sleeping entirely on the ground.[32]

Some theorize that, like numerous other organisms, hominins sometimes remained isolated in closed forested refugia (e.g., on the coast of East Africa) where they underwent intensive evolution, and then dispersed from these core areas to more marginal parts of Africa when environmental conditions were favorable.[33] Today, chimpanzees and gorillas are relic populations that live in much smaller parts of Africa than they used to. Unlike living apes, surviving hominins did not remain confined mostly to forested refugia but, instead, adapted to all kinds of environments around the globe, including a variety of forested habitats.[34]

It wasn't just that continuous closed forests began to fragment at the time hominins emerged. Big animals that browsed on trees, shrubs, and herbs, such as prehistoric elephants and black rhinoceroses, also began a steady decline in East Africa around 4.6 million years ago.[35] This has been attributed to the changing environmental conditions that caused tropical grasslands to expand at the expense of their vegetal sources of food.[36] The declining food supplies for large animals included trees that, presumably, would have provided suitable locations for our early ancestors' night nests. Thus, it appears that a decline in the number and density of available sleeping trees may have been associated with an overall trend toward cooling. If so, this could have been a significant factor in why hominin sleeping nests eventually went to ground – especially for those ancestors who eventually found themselves in more open habitats and grasslands.

In sum, although anthropologists argue vociferously about the precise nature of certain early hominin habitats, many now agree that our predecessors gradually shifted away from living primarily in dense humid tropical forests (similar to the habitats of apes) to occupying a variety of drier and less dense mosaic seasonal habitats that included mixtures of dry open forest, woodland, and open grasslands known as "savannas."[37] As they moved away from closed forests, somewhere down the line hominins shifted from sleeping in trees to sleeping on the ground. As we will see in the next chapter, this change had a profound impact on the evolution of hominin sleep, cognition, and sociality.

Why Walk?

From the previous chapter, we can assume that our earliest ancestors slept in trees at night but spent many of their daylight hours on the ground – resting, socializing, and traveling to find food and other resources – as living gorillas and chimpanzees do.[38] But what prompted some of them to begin moving on their hind limbs rather than mostly on all fours like living apes do when they are on the ground? And what would have caused this behavior to increase gradually over the course of the Botanic Age? Paleoanthropologists have long speculated that a primary benefit of walking upright was leaving the hands free to carry things, which seems reasonable. But what would a not-yet-bipedal ancestor need to carry? Probably not infants, because at the beginning of the Botanic Age hominin babies would have had the ability to cling securely to their traveling mothers, as living monkey and ape infants do, with little chance of falling off. Not food, because, like apes, these ancient ancestors would have eaten on the go rather than bringing food back to the dinner table.[39]

One reason why our ancestors started to walk upright more often when they were on the ground may have been related to their need to collect resources that were spread out across the terrain, including botanical materials for constructing terrestrial sleeping nests. Clues about when and how increased walking arose in hominins may be gleaned by studying the existing communities of great apes in which some individuals occasionally construct night nests on the ground, usually near or under the trees in which other group members sleep.[40] The highest percentage of ground nesting has been observed in a group of chimpanzees in the Nimba Mountains of West Africa, in which a startling 20 percent of nests are made on the ground, including nests for sleeping at night and for resting during the day. Primatologist Kathelijne Koops (figure 2.6) conducted genetic tests on shed hairs in ground nests and nearby tree nests to learn about who slept on the ground and how they were related to the individuals that slept above them in trees.[41] She found that night nests on the ground were made predominantly by males. Ground nesting was widespread, did

Figure 2.6. Dr. Kathelijne Koops taking it easy in a chimpanzee ground nest. Photo courtesy of Kathelijne Koops.

not seem to run in families, and likely occurred regularly in at least two different communities. Significantly, Koops concluded that since a fully terrestrial lifestyle was not necessary for regular ground nesting in these chimpanzees, it likely wasn't a prerequisite for the transition to ground sleeping in early hominins. The implication is that sleeping on the ground may have happened much earlier in hominin evolution than we believe! (Interested readers can find out more about Koops' research in her interview at the end of the book.)

Based on Koops' research, it seems likely that some individuals among our earliest ancestors also occasionally slept in ground nests at night, which eventually paved the way for the full transition to terrestrial sleeping. But why sleep on the ground when you can slumber more safely in a tree? For the most part, apes confine the little terrestrial sleeping they do to those rare areas where predators are not a threat. This has led to the assumption that if apes choose terrestrial sleeping sites when predators are not around, "terrestrial sites must be superior in some way."[42] Various suggestions have been offered about the kinds of factors, besides lack of predators, that might be associated with increased ground sleeping. The suggested variables include environmental conditions, habitat types, body size, social interactions, and quality of materials available for nest construction.[43] Although all of these things likely played some role, a good deal of evidence suggests that environmental conditions may have been especially important.

For example, one clever study of chimpanzees at a relatively open dry savanna site in Uganda estimated comfort levels associated with sleeping in ground nests versus tree nests by measuring and comparing the nests' relative windiness, temperature, humidity, and heat indices. What the researchers found was that the ground nests were significantly cooler, less thermally stressful, and more humid than the arboreal nests. In other words, since there were no predators around, chimps at this site likely shifted to terrestrial sleeping for the simple reason that it was cooler and less windy on the ground.[44]

Nesting Materials

Even though most living apes inhabit closed forests, a small number of chimpanzee populations, like the one in Uganda, live in hot, dry, open environments that generally resemble the habitats of many early hominins. These populations, often called "savanna chimpanzees," are not as well known as other chimpanzee populations because they have lower population densities, are spread out over larger home ranges, and are thus harder for primatologists to track and study.[45] Nevertheless, these communities are of keen interest because they provide clues about how hominins might have adapted to novel savanna landscapes after closed forested habitats fragmented following the Late Miocene Cooling. Significantly, savannas are characterized by low availability of suitable nesting trees and materials,[46] which may help explain why chimpanzees who live there sometimes build their sleeping nests on the ground.

Compared to arboreal nests that incorporate trunks and branches in addition to leafy matter, ground nests of apes are made from saplings,[47] nonwoody vegetation,[48] or both, and it is likely that this was also true for the ground nests our predecessors made. Unlike the tree nests that apes usually construct on the spot by pulling in and weaving surrounding vegetation, the materials for ground nests would not have been so readily at hand, particularly in the drier conditions of savannas that became increasingly common 2.7 to 3 million years ago during the Stone Age.[49] In such circumstances, individuals may have had to fetch botanical materials from further away and bring them to the nesting location, presumably by carrying them in their arms and hands. Increased reliance on ground nests thus might have been one of the triggers for the prolonged evolution of walking on two legs, especially since every builder would have needed new nest-building materials each day.

The idea that construction of ground nests may have been one of the behaviors that paved the way for an increase in bipedal activities and its associated physical adaptations (e.g., in the feet) is, by necessity, based on evidence that is circumstantial and cumulative. But happily, it also lends itself to a testable prediction – namely, that living wild apes that build ground nests frequently gather the necessary materials in their arms or hands and carry them bipedally on the ground to their construction sites. Regrettably, this idea is difficult to test because most reports analyze ape nests after their occupants have left them instead of providing eyewitness accounts of how they are built.

There are a few suggestive reports that, although not specific to ground nests, concern material collection for nest making by the one great ape that rarely comes to ground – namely, orangutans.[50] Orangutans are occasionally reported to collect nesting materials within trees and carry them to their nest-building sites, a behavior dubbed "leaf-carrying."[51] An early study noted that orangutans sometimes transported branches for making nests in their teeth, under an arm, in a foot, or between the shoulder and chin.[52] Another reported that an adolescent female repeatedly left the tree nest she was making to collect five or more long forked branches from elsewhere in the tree, which she placed on her neck, carried back to the nest, and used to line it.[53] Given that highly arboreal orangutans sometimes carry botanic materials to nest construction sites in these somewhat awkward manners (at least from the perspective of a striding biped), it is easy to imagine that apes capable of walking upright on the ground would be inclined to use their hands to collect and transport materials for the ground nests they made. Indeed, Jane Goodall discovered that a wild chimpanzee she named David Greybeard used this bipedal form of transportation when he stole armloads of bananas from her camp![54] The extent to which the African great apes walk bipedally while carrying vegetation in their arms to construct ground nests remains to be seen.

In any event, sleeping nests undoubtedly descended from the trees to the ground as hominin evolution progressed and baskets in the trees transformed into baskets on the ground. But the emergence of bipedalism and the gradual increase in terrestrial nests weren't the only things associated with the shift to ground living during the Botanic Age. As chapter 3 details, our predecessors' increasing habit of sleeping on the ground profoundly altered not only the quality of their sleep, but also their cognition.

3

Did You Make Your Nest This Morning?

YOUR BED IS, FOR ALL PRACTICAL purposes, a tree nest that's gone to ground. Although both human beds and ape tree nests are constructed to support one or two horizontally resting bodies (or more in human cultures with co-sleeping), there is a big difference between the way humans and apes manage their nests. Most apes make a nest in a new tree at the end of each day; humans usually return to the same bed, be it a Western-style mattress and bedding, a sleeping platform with hides for covers, or a hammock. This may be why many people make their beds in the morning, rather than at the end of the day as apes do. The most important function of both ape and human beds is the same, however: to enable a good night's sleep. Remarkably, the ape invention of deliberately constructing sleeping nests (baskets in the trees) didn't just kindle cognition about the physical world that enabled other inventions (the folk physics discussed in chapter 1). As detailed below, the more recent habit of building them on the ground had impacts on the evolution of sleep itself.

Hominins' transition to sleeping on the ground was not easy. Susanne Shultz, an expert on predator/prey interactions, notes that early hominins lacked natural defenses like large canines, big bodies, and fast running speeds, which would have made them vulnerable to terrestrial predators such as cats and large snakes, and also to raptors like crowned hawk eagles that target primates.[1] The fossil record confirms Shultz's suggestion that birds and leopards preyed on australopithecines (figure 3.1).[2] From observing other primates, she also speculates that early hominins may have developed strategies for dealing with predators such as mobbing or counterattacking them, possibly with sticks and rocks; using alarm calls to warn others of their presence; and finding secure places to sleep. Trees were obviously a safer place to sleep than the ground – and they still are, which is why great apes still make arboreal night nests. (You can read more about Shultz's ideas in an interview at the end of the book.)

In all great apes, adult males are considerably larger than adult females on average,[3] as is also true for living humans, albeit to a lesser degree. The difference in body size between the sexes in early hominins was also large. Because they were smaller, hominin females and youngsters would have been able to sleep in tree nests constructed from thin and low-weight-bearing branches that were high up and away from tree trunks, and thus less accessible to big predators such as pythons and leopards. As they evolved larger bodies over time, adult male hominins likely made their sleeping nests on lower sturdier branches or on the ground near tree trunks, similar to what some gorillas and chimpanzees have been reported to do.[4]

For this reason, it seems likely that males began the trend toward terrestrial sleeping, which increased as hominins became larger and dispersed to more open habitats, and the numbers of sleeping trees decreased in response to climatic changes. The lighter sex would not

Figure 3.1. Taung child killed by eagle. Illustration of a prehistoric eagle with its hominin child prey in its talons. This is the famous Taung (*Australopithecus africanus*) infant. *Australopithecus africanus* is one of several extinct species forming an early part of the hominin evolutionary tree. Illustration by Mauricio Anton/Science Photo Library.

have begun habitually sleeping on the ground until infants started to lose the ability to cling securely to their mothers while they climbed trees to make night nests (chapter 4). Meanwhile, adult males that slept terrestrially would have not only eliminated the risk of falling from nests that were too flimsy to support them, but they would have also been in good positions to prevent nocturnal predators (or amorous males[5]) from climbing up trees to get at sleeping females and youngsters.

If the australopithecines Lucy and Little Foot (chapter 2) are any indication, the complete shift to constructing night nests on the ground had not yet occurred by the end of the Botanic Age, at least in females. Instead, other fossils suggest that ground sleeping became the norm for both sexes only after habitual bipedalism was fully evolved in early *Homo* – perhaps around 1.5 million years ago as suggested by fossilized foot bones,[6] although some researchers estimate a more recent date of around 1 million years ago based on fossils of upper arm and shoulder bones in some representatives of *Homo*.[7] In any event, after hominins no longer slept in trees, and for the first time in prehistory, adults of both sexes would have slept with or near their offspring (often in the same nest), if only for safety's sake. This is still the norm for most non-Westernized[8] human cultures such as the Hadza in Tanzania.[9]

In other words, sleeping in groups was likely a necessary step for countering the pervasive threats from extremely dangerous nocturnal predators[10] – necessary but not sufficient, especially before hominins learned to collect, keep, and use fire to ward off prowling animals, which appears to have been a relatively recent achievement.[11] By the time fire did come around, however, hominins had successfully co-existed with nocturnal ground-dwelling predators for an extremely long time. One reason they were able to do so was because of physiological changes in how they slept. Serendipitously, these changes did much more than help early hominins ward off dangers of the night. But before we can get to all that, we need to delve a bit into the physiological underpinnings of modern sleep.

How We Sleep

Although you spend a substantial portion of your life sleeping, you may not have given much thought to how or why you do so. Perhaps it seems as if you simply go to bed, close your eyes, fall asleep, maybe dream, and then awaken. However, studies of the brains, eyes, and

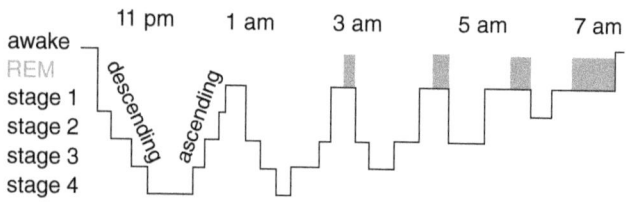

Figure 3.2. Typical sleep pattern for mature humans. After falling into a light sleep (stage 1), individuals descend into increasingly deeper stages of slow-wave sleep (stage 4 is the deepest), and then ascend the stages in reverse order. During the latter part of the night, stage 1 is followed by episodes of rapid eye movement sleep (REM, shaded blocks). Adapted from Buzsaki 2006.

muscles of sleeping people show that so much more goes on while sleep progresses during the night.[12] Sleep is composed of two basic states that alternate throughout the night: slow-wave sleep (SWS), in which the eyes are relatively still, and rapid eye movement (REM) sleep, which is accompanied by fast electrical oscillations in the brain (called EEG brain waves).[13] (See figure 3.2.)

While drifting off to sleep a person relaxes into a light SWS (stage 1) from which they can easily awaken, which has been described as a kind of "twilight zone between sleep and wakefulness."[14] They then sink into a deeper sleep (stage 2) from which it is harder to awaken, and finally glide into the deepest stages of SWS (3 and then 4, from which it is most difficult to awaken).[15] After reaching the deepest level of SWS, activity in the sleeper's brain ascends through the slow-wave stages in reverse order until it returns to stage 1. The four stages of SWS dominate the first part of the evening. Later in the night, stage 1 is followed by faster brain waves accompanied by REM.[16] Brain waves during REM appear most like those of awake people, which is why REM sleep is sometimes called "paradoxical sleep." People awaken naturally after the last REM sleep during the night – no alarm clocks needed.[17]

The brain performs different functions as the night progresses. Restorative housekeeping takes place during quiet SWS, which includes fixing connections between neurons, removing recently accumulated toxins, and securing fresh memories.[18] During REM sleep, new memories are transferred into more permanent storage, and people experience their most vibrant dreams but are inhibited from physical movements, except for their eyes and some spontaneous body twitches.[19] Some believe that such "sleep paralysis" helps prevent sleepers from acting out their dreams. REM sleep is conducive to learning and subconscious creativity (more on this below).

As any parent will tell you, sleep patterns change across the lifespan "with rapid cycling between sleep and wake states in infants, regular napping behavior in early childhood, and a consolidated nighttime sleep period in adults."[20] As a rule, very young infants are awake only 10 to 30 percent of the time and spend considerably more of their time in REM sleep than older people do.[21] Sleep is important for infants' development of language, which is likely a part of the reason they do so much of it.[22]

How long people sleep varies to some degree with weather, temperature, and natural light. Generally speaking, and contrary to the notion that humanity slept "better" before the invention of the lightbulb, people with and without electricity sleep, on average, approximately 6–7 hours a night[23] – 7 is considered healthiest.[24] (Interestingly, the widely held misconception that the "ideal" amount of sleep is 8 hours per night stems from the industrial revolution when laborers negotiated time off from factories for sleep.[25])

Sleep Patterns in Chimpanzees

To consider sleeping patterns that may have been present in early hominins before they became habitually terrestrial, we turn once again to our closest cousins, the chimpanzees. Although wild chimpanzees typically occupy their sleeping nests from dusk till dawn, their sleep is interrupted by bouts of "vocalizing, urinating, feeding, crop raiding, and socializing."[26] The ease and frequency with which they drift in and out of sleep is closer to those of people living in small-scale societies than in Westernized ones, as described below. Chimpanzees clock in at approximately 10–12 hours of sleep per night compared to the human average of 6–7 hours.[27] Chimpanzees are not only in bed longer and sleep more than people, but they also have a greater number of sleep cycles per night. Curiously, the one thing that does not seem to differ between chimpanzees and humans is the total amount of REM sleep accumulated per night, about 1 or 2 hours.[28] Humans simply consolidate their 1–2 hours of REM into longer stretches within fewer sleep cycles. This is important because, as we will see, anthropologists theorize that REM sleep was especially important for the emergence and evolution of higher cognition in early hominins.

Sleep in Traditional Societies

The belief that mature humans sleep during one uninterrupted period, usually at night, comes mostly from studies of people living in Westernized societies. Remarkably, this generalization breaks down in traditional (small-scale or preindustrial) societies[29] that lack electricity, live in nomadic bands or small isolated villages, and feed themselves by hunting or fishing, gathering plant foods, and (in some) tending family gardens.[30] Despite the fact that people in such societies are rapidly adopting Westernized technologies, their sleeping habits are of special interest because they differ from Westernized norms and could in some ways resemble sleeping patterns of prehistoric humans who lived under similar circumstances. The Westernized pattern of sleeping alone or with a partner on a heavily cushioned surface inside a roofed dwelling with solid walls during a single stretch of time is at odds with the sleeping arrangements observed among most traditional societies. For starters, infants in small-scale societies usually sleep with their parents and breastfeed during the night (known as "breastsleeping"), instead of sleeping separately (often in their own rooms), as is common in Westernized countries.[31] Other sleep habits are very fluid in traditional societies:[32]

> Neither !Kung nor Efe have bedtimes, so time of falling asleep varies widely within and among individuals. People stay up as long as something interesting – a conversation, music, dance – is happening ... then they go to sleep when they feel like it. Indeed, someone may go to sleep and get up later because they hear something going on ... Virtually no one is told to be quiet because others are sleeping, though people avoid unnecessary disturbance of sleepers. Additionally, no one, including children, is told to go to bed, and individuals of any age may nod off ... and fade in and out of sleep during night-time social activities ... In sum, sleep in these traditional societies is collective, and it occurs in social space.[33]

Despite sleep's social nature, people in small-scale societies commonly regard it as a potential time of danger from both physical and supernatural threats.[34] Among the Gebusi, for example, "In

Figure 3.3. A sleeping hut that was constructed by Hadza women (left) on a frame of flexible branches (right). Reproduced from Samson et al. 2017.

dreaming, spiritual life is activated, and spirits come to the dreamer who also enters their world. Deep sleep is considered risky because the sleeper's spirit may wander off too far and partially or wholly fail to return."[35]

Sleeping indoors may, to some degree, mitigate concerns about safety by contributing to a sense of security. Bedrooms differ, however, depending on the culture. Unlike industrialized societies, people in traditional cultures often use plant materials and techniques handed down from the Botanic Age to construct their shelters. Among the Hadza, for example, sleeping huts are made by several women who gather flexible branches that they bend and stick in the ground and crisscross with other branches to create an upside-down bowl frame, which is then stuffed with grasses[36] (see figure 3.3). Bedding materials (some of which are obtained through trade) may include animal hides, textile blankets, linen sacks, nets, or woven grass mats.[37] (Interestingly, before 200,000 years ago, people were placing grasses on top of ashes to create comfortable beds, often near hearths, at the warm back of Border Cave in South Africa.[38])

What differs between traditional and Westernized cultures is not so much the amount of sleep individuals get per night, but rather the time they spend in bed, which is generally one to two hours less in Westernized cultures than in traditional ones.[39] People in Westernized cultures pack their sleep into a relatively shortened time in bed because it is typically less fragmented by waking up and socializing than it is in small-scale societies.

Human Sleep: An Evolutionary Puzzle

Even though sleep is essential for physiological, cognitive, and emotional well-being, people sleep less than any of the hundreds of other species of primates. David Samson of the University of Toronto refers to this inconsistency as the "human sleep paradox."[40] Despite the decrease in total sleep, the accumulated REM sleep per night does not appear to have shortened during human evolution. The duration of individual bouts of REM simply increased as they became incorporated into fewer sleep cycles, which only adds to the enigma. People are extremely vulnerable during REM sleep because they cannot awaken easily, may be temporarily paralyzed, and are disconnected from what is going on in their environments. So why would individual episodes of REM consolidate into longer stretches?

Taking a cue from small-scale societies, Samson offers a "social sleep" explanation to help make sense of the sleep paradox: "In essence, by shortening human sleep duration ... ancestral humans had more time to develop skills, enhance knowledge, vie for mates, support offspring, and craft alliances."[41] It is likely that total time in REM sleep could be preserved in fewer sleep cycles because ground-sleeping humans eventually developed social and technological means to keep sleepers safe from dangerous predators. These safety measures likely included lookouts who remained awake and vigilant while others slept, as well as fire and shelters of some sort.[42]

However, the first hominins that transitioned to full terrestrial sleep would not yet have lived in settled communities, made sleeping huts, or used fire to ward off predators. Nocturnal predators would have been a *big* problem. Early hominins likely responded to this challenge by shortening the time they slept (compared to their ape ancestors),[43] sleeping in groups, and having sentinels. At some point, mothers who had difficulty climbing trees while carrying infants (more on this below) would have begun to make their night nests on the ground.

An increase in the number of ground-sleeping mothers and infants may have nudged others to join them, as suggested anecdotally by a group of captive chimpanzees who "nested on the ground to remain beside an elderly, ailing group member" who was unable to make an arboreal nest.[44] As Samson suggests, it is likely that social organization in hominins began to change as they bedded down on terra firma. Eventually, group living may have become more organized around a hierarchy of males who protected females and youngsters, as is the case for many semiterrestrial primates today, such as baboons and macaques.[45] But however the shift to ground sleeping developed over time, the duration of sleep eventually shortened and its neurological underpinnings changed in ways that helped not only our ancestors but also our modern selves to respond to fears, ponder the unknown, solve problems, learn, and make discoveries.

There's Something about REM Sleep

The fact that total REM sleep did not decrease with overall sleep time during hominin evolution suggests that there may be something very important about this particular stage of sleep. Various researchers have speculated that the vivid dreams that happen during REM have different functions than the mundane dreams, thoughts, and snippets of ideas that occur during SWS. REM sleep is thought to be important, not only for consolidating new memories (including learned procedures like how to ride a bicycle), but also for dealing with anxieties and threats, rehearsing possible future actions, solving problems, conceptualizing new discoveries, and inventing things.[46] Fortunately, rather than being homogenous, REM sleep contains substages in which the sleeper is more alert to the outside world.[47]

Memories and emotions are processed during the later stages of REM, and it is at this point that individuals do some of their most innovative and creative unconscious thinking.[48] Indeed, numerous inventions and discoveries reportedly came to their inventors during vivid dreams, such as the automatic sewing machine and the periodic table.[49] Similarly, musicians from shamans to classical composers have been inspired by dreams, including Richard Wagner, Igor Stravinsky, Giuseppe Tartini, Paul McCartney (who reportedly dreamed the melody for "Yesterday"), U2, Sting, and Billy Joel.[50] Sometimes music is not intrinsic to a musician's activity within the dream but instead feels as if it's coming through an outside channel. For example, Tartini wrote his famous violin sonata "The Devil's Trill" immediately after waking from a vivid dream in which the Devil played it on the violin with amazing virtuosity. Tartini, who believed that music could come from a supernatural gift, later said the intensity of the sonata he captured was not comparable to the Devil's.[51]

During most dreams, the dreamer has strong emotions and perceives a hallucinated and chaotic dream environment in which the flow of the narrative happens to (rather than is directed by) the dreamer. In such dreams, the person lacks self-awareness and does not realize the experience is a dream.[52] Although it is a relatively rare occurrence, people sometimes experience a state of lucid dreaming during REM sleep in which they become aware they are dreaming and can intentionally control their actions within the narrative – in other words, they have a dream within a dream that relies on highly evolved parts of the brain.[53] (Your dog may, indeed, be chasing a rabbit in his dream, but Fido is unlikely to be aware that he is, in fact, just dreaming.)

Dreams can occur at any point during sleep, although the most vivid ones are associated with REM sleep. Interestingly, not all lucid dreaming or creativity takes place in the REM state. Thomas Edison, who invented the phonograph and incandescent lightbulb among other things, identified a fleeting moment as he fell asleep in which he experienced insights about inventions. Apparently, it was his practice to hold balls in his hands as he drifted off to sleep so that he could capture productive ideas when he was awakened by dropping them. Edison's procedure has recently been tested, and it turns out he was correct. Those fascinating moments of fragmented perceptions that occur just as one drifts off to sleep (non-REM stage 1 in figure 3.2) are, in fact, associated with activity in the same neurological networks that are active during insightful thinking when awake. Falling into a deep sleep too soon, however, counteracts the potential effects of the "creative sweet spot within the sleep-onset period."[54] When one is working on a project or problem or needs to make an important decision, it is probably not a bad idea to "sleep on it" by subconsciously thinking about the problem as your brain drifts, preferably slowly, into sleep. Just have a pencil and paper ready to record your thoughts upon awakening.

Nighttime Anxieties and Dreams: An Evolutionary Perspective

The fossil record shows that there was a larger difference in the average body sizes of males and females in early hominins than we see in humans today. This is one reason why scholars speculate that males may have started the trend toward terrestrial sleeping, while females and youngsters continued sleeping in trees to avoid dangers from nocturnal terrestrial predators such as saber-toothed cats and prehistoric lions. Like some of today's large male gorillas, individual australopithecine males may have made nests on the ground at the base of trees where other more vulnerable individuals slept. Nonetheless, as suggested by evolutionary psychologist Richard Coss, females nesting in trees would still have been "vulnerable to nighttime attacks by leopards ascending from underneath them," while ground-nesting males would have been more "vulnerable to sideways attacks by leopards and other large, nonarboreal cats."[55] Coss speculated further that sleeping terrestrially may have changed the nature of male hominins' susceptibility to nighttime predators compared to females.

Coss wondered whether the pervasive nighttime anxieties of modern humans might reflect ancestral fears about predators. To explore this idea, he gave questionnaires to 140 American preschoolers (74 boys and 66 girls, ages 3–4) who had previously answered "yes" to the question, "When you were little, did you ever think there were scary things in your room?"[56] Follow-up questions asked whether the preschoolers thought the scary things were located above, at the side of, or below their beds. Neither sex identified scary things as being above their beds to any appreciable degree. However, more boys recalled that scary things were located at the side of their bed rather than below their bed. The results were reversed for girls.[57] When it came to naming the scary things, girls were chattier than boys. Individuals of both sexes identified dinosaurs, ghosts, snakes, spiders, monsters (the predominant category for both girls and boys), and cats/tigers (much more frequent in girls).[58]

Children may experience a particular kind of nightmare known as night terrors, which occur most often in one-to-five-year-old children and are typically accompanied by piercing screams, intense fear, inconsolability, and difficulty in waking up. One study observes that "the benefits of a child sleeping with a parent were likely manifold in the evolutionary environment in which humans evolved – most obvious is the decreased risk of predation of children if parents were nearby. Indeed, when school-aged children have any memory of a night terror they typically report indistinct recollections of threats (such as monsters, spiders, snakes, etc.) or fear of 'something' … that is going to get me."[59] This suggests night terrors may be an extreme response on the part of children to the fact that parents in Westernized cultures no longer sleep with them.

Dreams and nightmares have clearly played an important role in the history of religions the world over,[60] and people in traditional societies often interpret their dreams as evidence that beings such as ghosts are real. Researchers have found that non-Westernized people in many parts of the world "treat dreams as the sources of their religious ideas, including their concepts of their gods and other supernatural beings … It is likely that ancestral populations also treated them as such."[61] Although such dreams may entail reassuring visits from departed loved ones, others frequently involve encounters with unfriendly or demonic spirits.

Figure 3.4. *The Nightmare* by Henry Fuseli, 1781. Nightmares are experienced universally by humans. By Tulip Hysteria, CC BY 2.5.

Frightening dreams and nightmares are, in fact, a universal phenomenon that afflict people of all ages and walks of life (figure 3.4). Some scholars believe that having negative dreams in which one is chased or feels physically threatened is a mechanism for imagining potential real-life threats during sleep and simulating how to deal with them should they eventually occur. Because a large portion of such dreams incorporate scary animals, dangerous people, or threatening supernatural beings, advocates of this Threat Simulation Theory (TST) believe that the tendency for people to have such vivid dreams is a retention from the evolutionary past when animals and human enemies were more of a threat to life than they are today.[62]

Although not everyone concurs with the TST,[63] it seems reasonable to entertain it as a possible explanation for at least some of the dreams that people have. In fact, simulations that "allow people to 'preview' events and to 'prefeel' the pleasures and pains those events will produce" happen all the time in people who are awake and involve activation of a highly evolved "default" network in the brain.[64] (The default network includes parts of the brain that are active when one's mind is tuned to internal thoughts rather than to the external world.) Such prospection (as it is called) was and is adaptive because, "Alas, actually perceiving a bear is a potentially expensive way to learn about its adaptive significance, and thus evolution has provided us with a method for getting this information in advance of the encounter."[65] Because this part of the default network remains active when dreaming,[66] it appears that prospection continues even during sleep.

This is not to say that all, or even most, dreams reflect aspects of our evolutionary past (or a mismatch with it, such as parents no longer co-sleeping with infants). As we have seen, people tend to bring problems from their daily lives into the sleep mix. Many dreams are about the dreamer's sense of self and social relationships, which again relies heavily on the default network in the brain. Dreamers also process anxieties that stem from their current daily lives. What college student hasn't had one of the most common anxiety dreams – the examination dream, in which the dreamer is unprepared for a test that is about to be taken.[67]

Summing Up about Sleeping Down

Our earliest ancestors underwent dramatic behavioral and physical changes during the Botanic Age. For starters, they began transitioning from sleeping in carefully constructed arboreal night nests to sleeping on the ground. As we saw in chapter 2, this probably began to happen during the decrease in trees and shrubs that coincided with the Late Miocene Cooling around 7 to 5.4 million years ago.[68] As this shift slowly unfolded and continued into the Stone Age, hominins would have applied their considerable nest-making skills to making more of their sleeping nests on the ground, like today's great apes sometimes do. Instead of climbing trees and pulling in and weaving branches from their immediate surroundings to make sleeping nests, ground-sleeping hominins of both sexes would have collected botanical matter from the ground for constructing terrestrial beds. In fact, foraging for such materials and carrying them back to the nesting site may have been one reason for the increasing habit of walking on two legs (chapter 2).

The increased danger from predators associated with terrestrial sleeping was likely countered, at least to some extent, by the presence of adult males who protectively began to construct their own nests close to the ground nests of mothers and youngsters. This may have seeded the emergence of new forms of social life and group dynamics. Indeed, modern sleep schedules may be a vestige from when human ancestors made the final commitment to full terrestrial living – in other words, when all hominins were finally making their sleeping nests on the ground, despite its myriad dangers. Sleeping in groups and (much more recently) having a fire may have helped, but so would having lookouts who stayed up late ("owls") while others arose very early ("larks").

Changes associated with sleeping on the ground led to much more than a preference for sleeping safely in groups and having lookouts, however. The very nature of sleep changed. As our ancestors were transitioning to sleeping on the ground, the duration of a night's sleep shortened, which would have reduced the considerable risks from large nocturnal terrestrial predators (saber-toothed cats, hyenas, other hominins, etc.) and blood-sucking insects such as mosquitoes.[69] The architecture of sleep changed in such a way that the amount of all-important REM sleep did not decrease along with other kinds of sleep. With wariness about being vulnerable when asleep, many dreams were likely vivid and scary. At this stage hominins would not have invented language yet. They would not have been able to articulate their dreams, but they would have "felt" them to be real. Is it possible that such dreams contributed to the later emergence and enunciation of humanity's supernatural and religious beliefs, as some believe?[70]

Females did not join males in nesting on the ground because they'd finally gotten scary terrestrial predators under control. As you will recall, the fossil record shows that by the beginning of the Stone Age (around 3.5 million years ago) at least two species of australopithecine had feet with the kind of aligned big toes that are necessary for humanlike bipedalism but not so good for grasping. Although this was adaptive for walking, the eventual loss of prehensile feet in hominins and their babies, along with other developmental changes, meant that babies could no longer cling independently to their traveling and climbing mothers. This would have had harsh, even deadly, consequences as mothers with unweaned infants attempted to climb trees every afternoon to construct sleeping nests. One way that mothers would have eventually adapted to increasingly helpless babies, of course, was to start building their nests on the ground.

Besides prompting mothers to sleep terrestrially, the increased helplessness of babies likely sparked one of the most important inventions in prehistory – baby slings.[71] After all, early hominins would have inherited a proclivity for folk physics similar to that of the three living great apes. They would have intuited that helpless infants needed to be prevented from falling during their mothers' daily treks for food and the next place to sleep. It would not have been much of a cognitive leap to begin making little portable botanical nests to secure nursing infants to their mothers' bodies. As discussed in the next chapter, baby slings are used universally today and probably harken back to the first widely used portable tool that hominins invented after they split from chimpanzees.

4

From Tree Nests to Baby Carriers

SO FAR, WE HAVE ESTABLISHED THAT hominins were neither completely arboreal nor exclusively terrestrial for at least several million years after the ancestors of chimpanzees and humans split into their respective lineages. Instead of moving mostly on all fours when they were on the ground as their ancestors had, the hominins that emerged during the Botanic Age began to spend many of their daylight hours walking upright on terra firma. Although the earliest part of the hominin fossil record is fragmentary, it shows that even as their feet and legs were evolving for bipedal walking, our predecessors continued to have strong shoulders and grasping hands that allowed them to climb trees and construct night nests in which they could sleep in relative safety.[1] By the beginning of the Stone Age, only two species of australopithecines (Lucy's and Little Foot's) are known to have evolved aligned big toes that were more conducive to walking than to grasping. However, the strong shoulders and long mobile arms of these species suggest that they continued to spend a good deal of time in the trees. Accordingly, shoulders seem to have lost adaptations for arboreality more slowly than feet did, which is "consistent with a slow, progressive loss of arboreality" when one looks across the entire hominin fossil record.[2] In short, it took quite a long time for our ancestors to become completely bipedal in the way that we are.

In fact, it may have been as recently as 1 to 1.5 million years ago that prehistoric humans (*Homo erectus*) reached a point where none of the individuals in their communities routinely climbed trees – for the same reason that you don't. Do you have the shoulders and upper arm strength of an Olympic gymnast to pull your body higher and higher as you climb? Compare your hands to your feet. Could you grasp the trunk and branches of a tree and manipulate them with your feet and toes? Probably not. Unlike the (for all practical purposes) second pair of hands that apes have at the ends of their legs, human feet became weight-bearing rather than manipulative appendages. Even if you could climb a tree, you would, no doubt, find it impractical to do so every afternoon and then build a sleeping nest. This would be especially true if you were a nursing mother who needed to carry her infant up into the tree and keep it safe while making a nest. As we saw in chapters 2 and 3, although various environmental and ecological factors likely encouraged hominins to begin sleeping on the ground, mothers and infants (the genetic harbingers of the future) may have been among the last to join the terrestrial slumber party – and only after other individuals in the community were there to provide much-needed "safety in numbers."

Of all the higher primates (monkeys, apes, and humans), only human babies fail to develop an ability to cling with all fours to their mothers' bellies (figure 4.1), and to shift when they are somewhat older to riding on other parts of her body, like riding piggyback[3] (figure 4.2). This is not just because human feet lost their grasping

Figure 4.1. This baby monkey holds tight to its mother's body while the mother climbs and jumps between tree branches. By Ckpixel, CC BY-SA 4.0.

ability in the service of bipedalism. It is also because people lack an abundance of strong hair (fur) for babies to cling to, so even if human mothers adopted the slanted-back posture that knuckle-walking chimps and gorillas have when carrying their infants on the ground, their babies could slide right off their backs.[4] But long before the time of *Homo erectus*, babies losing their ability to cling presented a problem for our ancestors because mothers had to keep unweaned babies with them as they traveled during the day.

This is also true for great ape mothers, of course, but they do the best they can. Chimpanzee moms, for example, try to support their tiny infants by lending them an arm, leg, or lap ("groin pocket") and sometimes even carrying them. This is usually not necessary after infants are about two months old and have developed the reflexes and

Figure 4.2. Because this little chimpanzee can cling independently to its mother, she is free to use her hands and feet to climb the tree and make a sleeping nest. By Sekar B/Shutterstock.

grasping abilities needed to cling independently to their mothers. The first couple of months of a baby's life may be especially hazardous because falling off traveling mothers can be deadly. When Dutch behavioral scientist Frans Plooij studied wild chimpanzees with Jane Goodall, he observed a mother chimpanzee and her infant: "Madam Bee had raised two infants successfully when one of her arms was paralyzed during a presumed polio-epidemic … The two infants that were born afterwards died within a few months. I had the occasion to make observations on the first … Bee-hinde. Her body was full of wounds and scratches, so she must have fallen repeatedly. Whenever her mother moved about without supporting her, she whimpered and screamed continuously. Especially when support was needed most, such as during climbing, the mother failed to give it."[5]

This incident underscores an important point: after hominin babies lost the ability to cling and ride independently on their moms, the latter would have been stuck with the task of actively carrying them as they went about their routine of foraging for food and water. Some scientists have gone so far as to suggest that the final push for full bipedalism was in response to the necessity for mothers to free their arms and hands to carry their babies.[6]

When Did Hominins First Invent Baby Slings?

Any parent can verify that it is hard to get much done if you are carrying a baby in your arms. This is one reason why researchers Nancy Tanner and Adrienne Zihlman suggested decades ago that baby slings were among the very first tools, invented around the same time that australopithecines began to use natural containers such as ostrich-egg shells, pieces of bark, or animal skins to carry things – something that wild tool-using apes do not do. Further, these authors predicted, "eventually, perhaps, they developed means of combining the nest-making technique of interweaving strands of vegetation with the concept of carrying, to produce a netlike receptacle."[7] Indeed, it wouldn't have taken much of a lightbulb moment for the very first baby slings to be invented by prehistoric mothers, because the intuitive physics from a long heritage of weaving night nests, regarded by some as the first tools,[8] was already part of their nature. But how long ago did this happen, and in whom?

Fossil evidence from a skeleton of a tiny girl who died in Ethiopia when she was less than two-and-a-half years old[9] suggests that, instead of emerging in *Homo erectus* as some have suggested, baby slings probably came into play much earlier. The Dikika baby (an australopithecine assigned to *Australopithecus afarensis*) was capable of some grasping with her feet, but not nearly as much as ape babies are. If she was "still dependent on and perhaps often actively carried by adults,"[10] she and other unweaned infants in her species may have been especially susceptible to fatal falls at that point in time, several million years ago. Could *Australopithecus afarensis* mothers have used vines to secure babies to their bodies before climbing trees to make sleeping nests? Perhaps, because without something to prevent them from tumbling off their mothers, infants who could not remain attached on their own would have been vulnerable to severe natural selection, like the unfortunate little Bee-hinde. The first baby slings could have been as simple as a single, thin strap of botanic material wrapped around the mother's waist or slung over her shoulder.[11] Or were *Homo erectus* mothers the first to invent baby slings, as some believe?[12] Whenever they were first invented, once mothers started using baby slings, their babies would have been safer from natural selection (at least from fatal falls).

When considering the question of who first invented baby slings and when, it is worth remembering that *Australopithecus afarensis'* predecessors had nearly three million years to hone their intuitions about the physical world and to improve their manufacture of botanic tools. The oldest (very primitive) stone tools found so far are from a site in Kenya that dates to the same time as the Dikika baby (about 3.3 million years ago[13]) and cut-marked bones from around the same time were found at the Dikika site in neighboring

Ethiopia.[14] Together, these discoveries suggest that *Australopithecus afarensis* was well on the way to making simple stone tools (see chapter 5). It does not seem too much of a stretch, then, to suggest that by the time these australopithecines began to appear in the fossil record (3.7 million years ago) they had also figured out that, just as sleeping nests prevented adults and youngsters from falling out of trees, weaving and securing infants in little wearable nests would prevent them from dropping off their mothers. In fact, it is not too farfetched to speculate that baby slings themselves may have contributed to the emergence of full-fledged bipedalism by keeping babies that were developing bipedal (nongrasping) feet in the gene pool!

Alas, we may never know for sure who invented the first baby slings because the slings were almost certainly made of botanic matter that would not have fossilized and been preserved over deep time (in the next chapter we will see their first traces in the archaeological record). Nonetheless, if baby slings were invented millions of years ago for the pressing reasons suggested above, then one would not be surprised to see slings made of natural materials or textiles used widely by the descendants of their inventors – including people in non-Westernized societies. But are they?

Baby Slings around the World

A comparison of baby-carrying devices in small-scale subsistence societies can shed light on the first baby slings that were invented millions of years ago and how they became modified when hominins eventually migrated out of Africa. Despite the fact that the people living in them were biologically just as modern as humans everywhere else, ethnographic reports on these ethnographic-period societies are immensely valuable for thinking about early evolution because the modern living conditions and lifestyles in such societies more closely resemble those of early foraging and gathering hominins than those of Westernized societies.[15] Because these cultures are rapidly disappearing, much (but not all)[16] of the relevant research comes from anthropologists before 1960.

In a classic paper, American sociologist and anthropologist John Wesley Mayhew Whiting compiled information from a "world tour through the ethnographic reports" of baby-carrying devices that is not only fascinating, but also remains illuminating.[17] Since our early ancestors lived exclusively in Africa during the Botanic Age and much of the Stone Age, this seems like a good place to start. Among the !Kung San of the Kalahari Desert in Botswana in southern Africa, wakeful babies were held sitting or standing in their mother's (or another caretaker's) lap as she sat on the ground, or they were carried in a baby sling. The slings permitted infants to move their arms and legs freely and have skin-to-skin contact with their mothers. Baby slings also allowed toddlers ready access to mothers' uncovered breasts, so they could feed whenever they liked.[18] Slings typically secured babies to their mother's side (hip), which allowed them to see people and objects, more or less, from her point of view. Infants that fell asleep on someone's lap or in their arms were usually placed on a cloth next to the mother.

In other African societies, babies typically slept at night next to their mothers, often on a mat or rug, which allowed comforting skin-to-skin contact and nursing. In some groups in the Sudanian savanna, special baskets for napping and resting were made from pliable branches. After the first few months of life, these infants spent a great deal of time in body contact with someone, usually the mother, although helpers pitched in to a greater or lesser extent depending on the culture. These babies were usually carried in slings or shawls on the backs of their mothers or others. "Before the advent of Christian missionaries, most infants in sub-Saharan Africa wore no clothes, and the mother, except for the carrying shawl, was topless. Thus while being carried the infant was in skin contact with the mother."[19]

When Whiting turned his attention to Eurasia, he found that infants were rarely carried in slings or shawls. Instead, most of them were swaddled and placed in transportable cradles that could be rocked to comfort them and put them to sleep. (In sling cultures, this was achieved by carrying the infant on caregivers' backs as they walked or did domestic chores.) In some of the island societies in the Indian and Pacific Oceans, babies were more likely to be carried on the hip than the back (as in Africa), or simply transported in caregivers' arms. In New Guinea, a net bag was used as a carrying device worn over the mother's back and also as a kind of portable baby hammock to hold stationary infants. For example, the net baby sling from Papua New Guinea shown in figure 4.3 is perfect for a hot climate and practical for hanging under houses allowing babies to rock to sleep.[20] "Thus in general infants are cared for in the insular Pacific in much the same way as they are in Africa. They are in skin-to-skin contact with their caretaker most of the day and night."[21] Infant care in North America, on the other hand, was similar to that in Eurasia (i.e., cradles); south of the Rio Grande, cradle boards were, again, replaced with slings, shawls, or simply arms, although infants rested in cradles.

What this boils down to is that hot and mild winter climates are associated with baby slings and infants sharing beds or mats with their mothers, whereas cool or cold winters are linked with more infants sleeping separately in cradles or cribs (see figure 4.4). "In cold climates infants tend to be carried in a cradle, swaddled, and put in a

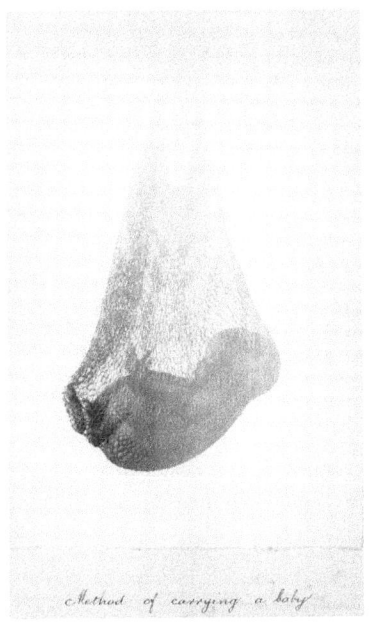

Figure 4.3. Net baby slings are perfect for a hot climate. Photograph taken in Papua New Guinea between 1901 and 1910 by Australian Anglican lay missionary Percy John Money. With kind permission from the Mitchell Library, State Library of New South Wales.

cradle to rest and nap during the day and to sleep at night. In warm climates they are usually carried in a sling or shawl, often nap on their caretaker's back, sleep next to their mother at night, and are clothed lightly or not at all."[22] It seems, then, that wooden cradles only came around after human ancestors migrated from Africa to colder northern regions. Less complicated baby slings would have been invented earlier in tropical Africa, and "the sling or shawl is such a simple idea that it may well have been reinvented frequently."[23] It follows that simple woven baby slings that allowed close contact between mothers (or other carriers) and infants were probably the norm for the majority of prehistory, which unfolded largely in tropical parts of Africa before some hominins started traveling to less hospitable climes.[24]

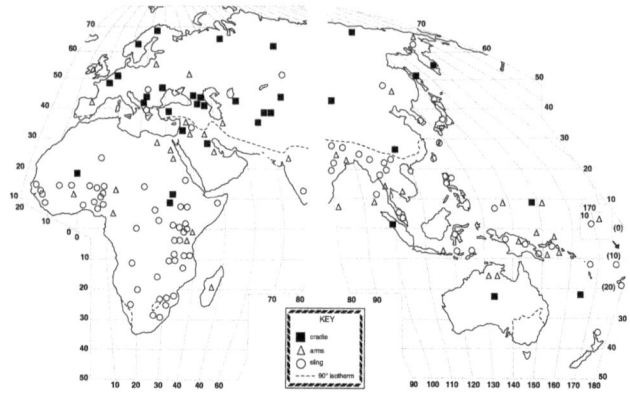

Figure 4.4. The use of slings (circles), cradles (filled squares), or simply arms (triangles) reported in the ethnographic literature. Warmer climes are associated with baby slings, colder ones with cradles. Reproduced from Whiting 1981.

People began to arrive in the Americas relatively recently, perhaps around 25,000 to 30,000 years ago,[25] having migrated from cold parts of Asia.[26] These immigrants brought their babies along and would have used baby-carrying devices.[27] Because infants in colder climates were usually carried in cradles,[28] it is not surprising that a heavy reliance on these was brought to the Americas.

Cradle use in the Americas is illustrated by the research of Alexandra Greenwald on ethnographic-period basketry cradles in Central California.[29] Native American groups here used botanic materials to weave two basic types of cradles – sitting and lying, each characterized by regional variations. Women were primarily responsible for making cradles, and for collecting the plant materials for weaving them. In deep sitting cradles, infants were diapered with soft plants like moss, swaddled, and strapped in a seated position with sinew, buckskin, or string (figure 4.5). In lying cradles, infants' bodies were

Figure 4.5. Woman with child in deep sitting cradle. Photograph by H.W. Henshaw, 1892. With kind permission from the Grace Hudson Museum and Sun House, City of Ukiah, California.

extended and attached more or less vertically to caregivers' backs, but could be placed in other positions when the cradle was removed from the adult's back. Greenwald notes that mothers took their nursing infants with them when they foraged daily for food: Cradles "were carried by the mother ... and then propped against a tree or rock, hung from or tied to a tree, or ... stuck vertically in the ground, enabling the infant to observe his or her mother's activities while in an upright position and being close at hand for nursing."[30] Greenwald's research beautifully illustrates the continued manufacture and reliance on woven baby carriers into historic times – an important legacy handed down from ancient predecessors who lived during the Botanic Age.

Although baby carriers in one form or another continue to be used in much of the world,[31] they were largely replaced by baby carriages (perambulators) in Western Europe beginning in the early eighteenth century.[32] Needless to say, this decreased how much the infant was held or carried, and infants spent more time in horizontal positions than vertical ones. Interestingly, infants in baby carriages or prams may have experienced "fewer motor challenges, less tactile and vestibular stimulation, less direct maternal contact … compared to young infants in hunter-gatherer societies" and the resurgence in the use of baby carriers and slings that began in the late twentieth century "may be a partial return to ancestral patterns."[33]

Man the Protector, Woman the Porter

As Tanner and Zihlman noted nearly 50 years ago, "a sling for carrying an infant less able to cling than chimpanzee infants could also serve as a gathering bag."[34] This is significant because devices for carrying resources were fundamentally important for the emergence of a gathering lifestyle in early hominins. In fact, bags, sacks, baskets, and other carryalls continue to be used universally to this day.[35] Baby slings probably came first because, from the start, mothers would have had to keep nursing infants on board while they traveled, just as ape mothers do.[36] Although wild chimpanzees live in gregarious social communities, sleep arboreally, and forage for food on the ground during the day, chimpanzees and other apes never evolved the habit of making and using containers to collect and transport items.[37] Instead of collecting food to share, chimpanzees feed as they go, which would have been the case for hominins when they first began sleeping on the ground. Once the shift to collective ground sleeping was underway, it is logical to assume that small groups of foraging hominins would have reconnected before bedding down at a central place to sleep, as some savanna baboons do to protect their communities from nocturnal predators. Thus, according to the Central Place Provisioning theory of the late Frank Marlowe, at first hominins simply reconnected to sleep and only later began collecting food to bring back to a central place to share with others, as was (and still is) typical with contemporary human foragers.[38] Initially, it is unlikely these places were permanent, since the earliest hominins are thought to have lived in small nomadic or seminomadic groups that moved in search of better resources when circumstances warranted. In any event, the purpose of baby slings probably broadened to include carrying other things when hominins began to forage for resources to bring back to their groups.

If you are fortunate enough to travel to places like Addis Ababa, Ethiopia, you will see many women who carry loads including water, firewood, produce, and other goods and wares – with or without babies riding along (figure 4.6). These loads are usually carried with the help of devices that attach them to women's fronts, backs, hips, across their arms and shoulders, or even on their heads. In general, it is the so-called "weaker sex" that does the heavy carrying, and for good reason. As shown in experiments carried out by Cara Wall-Scheffler and her colleagues, women are built physiologically to carry heavy loads, as they must do when they gestate growing fetuses and continue to carry them after they are born.[39] Although it is obvious that both sexes of early hominins carried things, Wall-Scheffler suspects there was eventually an evolutionary shift that conferred a

Figure 4.6. A woman carrying not only a baby, but other items as well. Photo by Dean Falk.

physiological ability for carrying heavy loads on women in particular, partly because women's adeptness at doing so seems to be universal. (You can find more details in the interview with Wall-Scheffler at the end of the book.)

The fact that women in many, if not all, non-Westernized societies typically carry most of the heavy loads is well documented by anthropologists. Unfortunately, this has sometimes been interpreted to mean that men forced women to become "pack mules." For example, a report from 1933 noted that "the accounts of travellers, who usually have little insight into native mentality, depict the women of Australian tribes as miserable slaves, as beasts of burden driven by the

Returning from the gardens

Figure 4.7. Man with spear in Papua New Guinea walking ahead of a woman bearing a heavy load. Photograph taken before 1910 by Australian Anglican lay missionary Percy John Money, who labeled it "man the protector, woman the porter." With kind permission from the Mitchell Library, State Library of New South Wales.

cudgel of their lord and master."[40] Not all interpretations from the early part of the twentieth century were so harsh on men, however. The photograph above that was taken in Papua New Guinea circa 1910 (figure 4.7) illustrates a woman walking behind a man who is carrying a spear. The woman leans over in order to carry what appears to be an extremely heavy load in addition to the baby that rides on her shoulders. Her backload includes a filled bag, a bundle of wood, and what appears to be a calabash that probably contained water. Although some might be tempted to interpret this photograph as validation of the "beast of burden" comment, the photographer (Anglican lay missionary Percy John Money) may have thought otherwise when he labeled the image "man the protector, woman the porter." Indeed, Money's labeled photograph encapsulates the idea that, once they started collecting resources to share, males probably provided more protection to groups than females did, while females were more likely to use slings and other devices to carry not only infants, but also objects such as food, much of which would have been botanical.

In chapter 5, we will see how the roles of females and males became even more differentiated as hominins evolved, finally invented stone tools, and eventually took up serious hunting.

5

First Came Wood, Then Came Stone

PREHISTORIC BABY CARRIERS WOVEN FROM PLANT matter did not fossilize, so the earliest record of their use is a picture etched around 15,000 years ago on a piece of engraved slate from the famous archaeological site of Gönnersdorf, Germany (figure 5.1).[1] Like many other depictions of females at the site, the four women that appear on this piece are highly stylized and lack heads. The women face to the right and the second female from the right is bent over underneath the weight of a cradle she carries on her back. The cradle (rather than a sling) was fitting for Germany's cold winters (chapter 4) and the infant who occupied it faced backward. Interestingly, the stooped posture of this mother resembles that of the woman from Papua New Guinea shown in figure 4.7.

Although the archaeological record of baby carriers is extremely recent because they did not fossilize, indirect evidence indicates that hominins were weaving textiles long before the first recorded baby slings. In fact, Helen Anderson of the British Museum in London

Figure 5.1. Engraved slate from the Gönnersdorf site in Germany, which portrays four stylized females who lack heads. Note that the baby (B) in the cradle (C) carried on the back of the second woman from the right faces backward. Adapted from Don Hitchcock 2015.

theorizes that basketry is a worldwide phenomenon that arose early during evolution and contributed to hominins' emerging understanding of numbers, patterns, and structures. Because woven textiles are very poorly preserved, Anderson examines geometric patterns associated with weaving that were imprinted or incised onto materials such as shells and bone, which "may indicate the transference of patterns and designs emanating from cordage, thread, nets, traps, and woven containers."[2] For example, some of the oldest geometric patterns were incised on pieces of ocher and engraved on stones from 100,000 to 60,000 years ago.[3]

Anderson notes that geometric patterns have shown up even earlier in the archaeological record. (See the interview with Anderson at the end of the book.) Astoundingly, around half a million years ago in Java, a *Homo erectus* individual engraved a shell from a mussel with a simple pattern that included parallel lines and zigzags. This "was considerably older than the oldest geometric engravings described so far."[4] At approximately the same time, but thousands of miles away, Wonderwerk Cave in South Africa yielded around 20 incised slabs (mobiliary art) "that often feature well-spaced parallel curved lines ... from stratified cave contexts extending back to ~0.5 Myr ago."[5] Although these artifacts existed long before the famous cave and rock art that hominins eventually produced, it is telling that the earliest forms of rock art consisted of the kinds of geometric patterns found in weaving rather than representational figures (the same is universally true of children's art) – a mysterious phenomenon known as the "geometric enigma." As Anderson has theorized, there is clearly something very basic about the perception and depiction of geometric patterns in the human nervous system.[6]

Other evidence for textiles is more direct. For example, a 45,000-year-old fragment of three-ply cord made from inner bark fibers was recovered from a site in France. This is significant because "twisted fibres provide the basis for clothing, rope, bags, nets, mats, boats, etc. which, once discovered, would have been an indispensable part of daily life. Understanding and use of twisted fibres implies the set of complex multi-component technology as well as a mathematical understanding of pairs, sets, and numbers."[7] Along similar lines, 30,000-year-old wild flax fibers were recovered in a cave in the Republic of Georgia that may have been used to make cords, haft stone tools, weave baskets, and sew.[8]

Although impressions on ceramics show that twining has been important in the Americas for at least 11,000 years, archaeologists have not always realized their significance. As archaeologist Alice Kehoe observed over two decades ago, "standard practice in American

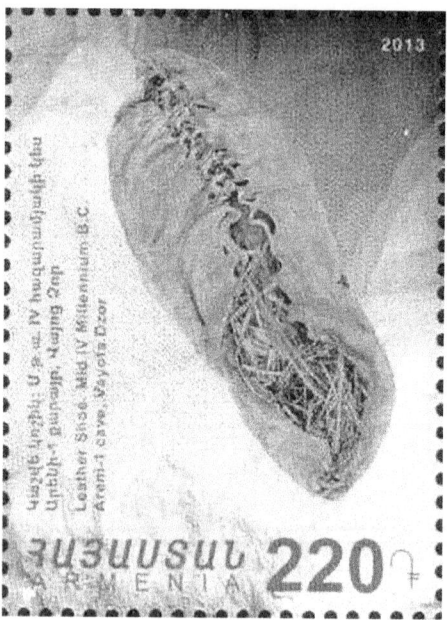

Figure 5.2. Commemorative stamp from Armenia (2013) showing the earliest known leather shoe (~5,500 years BP).
By Post of Armenia.

archaeology has been to give fabric impression on ceramics one sentence while devoting literally reams of paper to details of lithic artifacts ... [There is] no inkling that the fabrics are artifacts in their own right, available for study on hundreds of thousands of sherds. Emperors without clothes, authoritative archaeologists march on blind to the cultural information woven into fabrics."[9]

Fortunately, other discoveries of textiles have been more readily appreciated. For example, a 6,000-year-old skirt fragment (made of reeds) and the earliest known leather shoe (with laces) were commemorated on Armenian postage stamps (e.g., figure 5.2).

Figure 5.3. Reconstruction of shelters and tree canoes near Långbergen, Saltvik, Åland (Finland), as they may have appeared in the Stone Age, about 6,000 years ago. By Erik Wannee, CC BY-SA 3.0.

Archaeologists have also recovered remains of skillfully made wooden artifacts, including shelters and tree canoes that were crafted in Finland around 6,000 years ago (figure 5.3). The world's oldest known wooden wheel and axle were made around 5,000 years ago and eventually found on a floodplain in central Slovenia (figure 5.4).[10] Cradles, canoes, and wheels were important inventions that were (at least initially) made from vegetal matter. These derivatives from the Botanic Age also had something in common at a conceptual level. Each one was used to transport things from point A to point B – babies, other people, animals, agricultural produce, and more. Although wild apes sometimes use rocks to crack open nuts and have been known to modify branches to fish for insects, they do not make tools to transport things. The ability to conceptualize and invent vehicles, which is essentially what these artifacts are, emerged

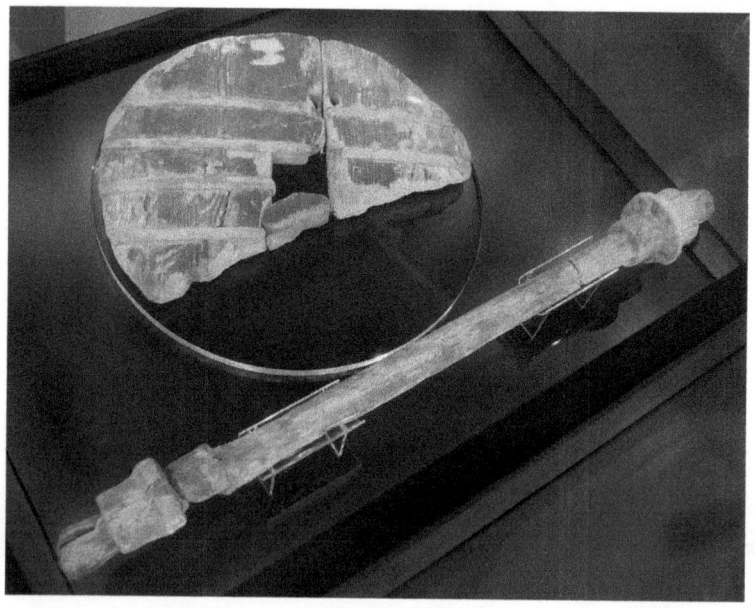

Figure 5.4. Oldest wheel yet discovered with axle from Ljubljana Marshes in Slovenia, which may date to around 5,000 years ago. By Petar Milošević, CC BY-SA 4.0.

in hominins after they diverged from chimpanzees. As we will see, botanic inventions may give stone tools a run for their money when it comes to identifying cognitive advances. But first, we need to examine why paleoanthropologists have given stone tools enormous priority over botanic ones when thinking about the emergence of advanced cognition, and then ask whether this bias makes sense.

The Crafting of Stone Tools

Any hand-sized rock can be used as a tool. Great apes sometimes throw them at others or use them to crack nuts. But actually making a tool by deliberately modifying a stone is something else entirely. The oldest recognized stone tools made by early hominins are simple flakes from a 3.3-million-year-old archaeological site in Kenya called Lomekwi.[11] These "unretouched" flakes hardly look like tools to a non-archaeologist. Instead, much like the items known as Oldowan "pebble tools" that began to appear around 2.8 million years ago,[12] they resemble rocks you might have kicked home from school as a child. Despite their simple appearance, however, these tools were deliberately made by using rocks to hit or "knap" flakes off larger stones called cores. This process created sharp-edged flakes that could be used to scrape meat off bones and chop or cut other items including botanical matter. Somewhat earlier but less direct evidence for stone tools comes from the Dikika research area in Ethiopia that is associated with Lucy's species (*Australopithecus afarensis*). Approximately 3.4-million-year-old cut and percussion marks were inflicted there on bones from hoofed mammals (ungulates), presumably when meat was removed from carcasses and bones cracked to obtain marrow.[13] Because of the evidence from Kenya and Ethiopia, this book places the beginning of the Stone Age at around 3.5 million years ago. The boundary between the Botanic and Stone Ages is likely to change, however, if and when even older stone tools are discovered (a good bet).

Given that a record of cultural artifacts is entirely lacking for the first three million years of our hominin heritage, and that the tools that eventually dominated the archaeological record were made almost exclusively from rocks, it is not surprising that some scholars associate evolutionary "cognitive leaps" with stone tools. Some, for example, have specifically focused on the emergence of the mental capacity to fashion symmetrical stone tools known as Acheulean bifacial hand axes, which *Homo erectus* began making along with other Acheulean tools around 1.8 million years ago (figure 5.5).[14] In fact, scholars Frederick Coolidge and Thomas Wynn identify these hand

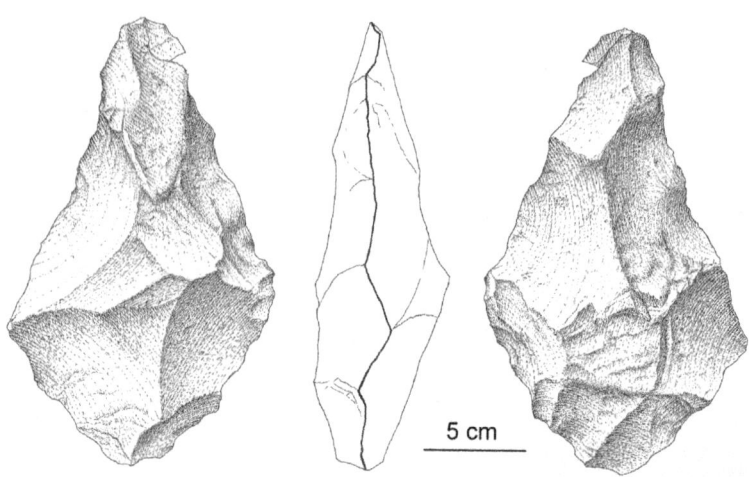

Figure 5.5. One of the earliest Acheulean hand axes on record (around 1.5 million years old), from Peninj, Tanzania. Reproduced from Diez-Martín et al. 2018.

axes as a hallmark of the "first major cognitive leap" in hominins because their production required more planning, sequential steps, memory, and physical coordination than the much smaller and simpler stone tools that started to appear in the archaeological record at the beginning of the Stone Age.[15] Coolidge and Wynn speculate that these bifaces were probably the first tools that existed in the minds of their makers *as tools* and, further, that the idea of shaping stone tools to be symmetrical might have resulted from creative insights during the dreams of hominins whose sleep patterns (and brains) had been modified in the course of shifting to sleeping on the ground (see chapter 3).[16] Thus, "the first apparent evolutionary development in cognition well beyond the ape range occurred with the advent of *Homo erectus* (upright man) ... along with a *lithic* (stone) technology known as Acheulean ... The hallmark of this technology is the handaxe ... made by trimming around the margins of a large flake to produce a sinuous cutting edge. In doing so, they also imposed a bilateral symmetry on the tool."[17] Apparently, when it comes to assessing the relative intelligence of our early relatives, rocks reigned and *Homo erectus* was the man.[18]

But how were these hand axes used? Paradoxically, less is known about their functions than about those of earlier stone tools. Unlike the simpler tools, Acheulean hand axes were not usually found with animal remains associated with butchery.[19] Instead, archaeologists have speculated that bifacial hand axes may have been heavy-duty multipurpose tools used to cut animal carcasses, scrape hides, and hack wood. One of the first studies that specified how early hominins likely used stone tools to process botanical material showed that Acheulean hand axes from the 1.7 to 1.4-million-year-old Peninj site in Tanzania (figure 5.5) were unambiguously used for woodworking.[20] Microscopic plant structures made of silica (phytoliths) and mineralized vegetable fibers were preserved on the internal working edges of two hand axes. Residues were also found on a flake that indicated it had probably been used to remove the outside surfaces of branches. The researchers concluded that the hominins at Peninj used the hand axes to make digging tools and spears from the hardwood *Acacia*.[21] Based on observations of modern humans, the authors further surmised that the vegetable fibers on the hand axes may have been remains of plants used to protect the hands of woodworkers. An implication of this study was that, in addition to other wooden tools, hominins may have produced rudimentary wooden spears much earlier in the Stone Age than previously believed.[22]

In the 20-some years since this study was published, additional evidence has accumulated that suggests stone tools were, indeed, very important for early hominins' gathering and processing of botanic resources. For example, a use-wear analysis of the edges of 62 Oldowan tools from the approximately two-million-year-old site of Kanjera South in Kenya found that 30 percent had been employed to process animal tissue while 70 percent were used on botanic matter such as wood, dirt-covered underground plants (tubers or bulbs), stems, and grass-like plants (sedges). The authors speculate that, in addition to harvesting seeds and plant food, the "cutting of grasses, sedges or reeds" may have also been related to "a non-subsistence related task (e.g., production of 'twine,' *simple carrying devices*, or bedding)" (italics mine).[23]

The earliest known hominin tools, to date, are the 3.3-million-year-old flakes from Lomekwi, Kenya. These flakes were relatively large, which is interesting in light of recent experiments by Rebecca Gürbüz that show larger flakes are more efficient for woodworking compared to smaller ones, which led her to conclude that "woodworking, just as much as butchery, might be a motivating factor in the appearance of this earliest-claimed instance of stone flake production."[24] (Interested readers can learn about these experiments in the interview with Dr. Gürbüz at the end of the book.) If Gürbüz's research and interpretation of the Acheulean hand axes from the Peninj site are any indication, we must seriously consider that the hallmark of hominins' "first major cognitive leap" may have emerged primarily as a tool for refining wood tools.[25]

The implication that Acheulean hand axes were the first sophisticated tools is far from perfect, however. The presence of complex stone tools in the archaeological record doesn't mean that the ability to make them "leaped" onto the scene or that equally sophisticated tools weren't being made simultaneously (or earlier) from other, more supple, plant materials. It is not surprising that there is no record of artifacts from the extended period of tool use before our predecessors figured out how to batter rocks into useful shapes, because organic matter is much less likely to survive than stone.[26] The thing to keep in mind is that the human toolbox accumulated and expanded over time with the addition of new inventions to earlier ones, which often continued to be made and used. (Just because you use a computer doesn't mean you don't also use a hammer upon occasion.)

Another caveat is that archaeologists sometimes assume absence of cognitive sophistication when material evidence for it is lacking. However, the supposed link between early hominin material culture and cognition has recently been reconsidered in light of contemporary hunter-gatherers' use of materials and found wanting: "Contemporary foragers are *just as cognitively sophisticated* as other contemporary human populations. Yet, even despite access to metals and plastics, alongside extensive exchange with neighboring agricultural groups in goods and ideas, many have artefact sets smaller and less elaborate than those associated with Upper Palaeolithic Europe."[27]

Accumulating research strongly suggests the earliest stone tools were important for making wood tools, but this has not yet had much of an impact on the field. Why do many anthropologists continue to emphasize the importance of the earliest stone tools for hunting and butchery while more or less ignoring their potentially significant role in making wooden tools? Since we presume that early hominins were at least as capable of making and using tools as living chimpanzees, it seems worthwhile to explore when and how chimps use and modify wood and rocks as tools.

Chimpanzee-the-Hunter

Recall that, although a few wild communities of chimpanzees in West Africa use rocks to crack open nuts, all populations of chimpanzees employ sticks as the equivalent of Swiss Army knives, which may be used as diggers, pokers, crude missiles, spoons, levers, clubs, insect fishing rods, and so on (chapter 1). And all great apes use sticks (branches) to make contemporary versions of the primordial stationary tool – the sleeping nest. Leaves may be employed to sop up drinking water or remove sticky fruit juice from the hands and face, and rocks are sometimes thrown to drive away other chimpanzees or animals.

The main reason chimpanzees use tools, however, is to obtain food. In fact, scholars have long theorized that apes (and a few monkey species) evolved the intelligent use of tools to extract particularly desirable seasonal foods such as insects, honey, underground tubers, seeds, fruits, and nuts from embedded locations.[28] In addition to this "extractive foraging," all chimpanzee populations across Africa, including bonobos,[29] hunt vertebrates and sometimes steal carcasses from other animals such as baboons. Hunting occurs much less frequently among chimpanzees than extractive foraging and, unlike foraging, rarely entails the use of tools.[30]

Chimpanzees prefer to hunt arboreal primates and are especially eager to capture immature red colobus monkeys.[31] Their next most frequent targets are terrestrial animals like young bushbuck, bushpigs, and duikers (a type of antelope). Rodents, nestlings, and eggs are also fair game. They will occasionally kill (and sometimes cannibalize) each other, and there have even been rare observations of chimpanzees seizing and killing human babies.[32] But generally chimpanzees favor small, immature prey, and they avoid animals that are too big to be safely captured like large baboons and ungulates.[33]

Although males (and, rarely, females) sometimes hunt alone, chimpanzees typically hunt arboreal monkeys in groups that consist of mostly adult and adolescent males. Depending on the community, individuals may or may not assume different roles that include driving, ambushing, and finally chasing monkeys to make kills.

Although hunts in most chimpanzee communities are not this coordinated, the basics of hunting monkeys seem to be similar across populations: A group of hunters spots and herds a group of arboreal monkeys (initially following underneath them along the ground), one or more chimpanzees ascend the trees to chase a target (e.g., a mother with an infant) and, with luck (for the chimpanzees, not the monkeys), a monkey may be grabbed and killed. The meat from a group hunt is normally consumed by the successful hunter who shares parts of the carcass (perhaps reluctantly) with others.

Hunting arboreal monkeys is physically demanding and usually undertaken by adult males who make the majority of kills. Accordingly, observations of red colobus hunts at sites in Tanzania and Uganda revealed that females engaged in less risky hunting by focusing less on monkeys and more on sedentary terrestrial prey like bushbuck or bushpigs.[34] Bushpigs are nocturnal animals that remain relatively still and hidden in dense thickets during the day. If chimpanzees see or hear pigs, they check the undergrowth carefully. Sometimes a piglet is found and seized before adult bushpigs can come to its defense (which they do vigorously). Females traveling with their family or with a few other females are more successful at obtaining and keeping piglets, bushbuck fawns, and even colobus monkeys than when they travel with a group that includes males.[35]

Females' preference for less risky terrestrial prey is consistent with the noticeable differences in the bodies of male and female chimpanzees (sexual dimorphism). Males grow up to be considerably bigger and stronger than females, and they have larger canines that they use in aggressive attacks. Adult females spend a significant portion of their lives gestating or carrying newborns and infants, which likely adds to the difficulty of chasing monkeys through the canopy and then capturing and killing them.[36] In other words, when it comes to hunting, physical differences between male and female chimpanzees appear to be associated with a sexual division of labor. Males pursue the bulk of the fast-moving arboreal monkeys; females tend to go for more sedentary or slower-moving terrestrial animals, although they have been known to hunt red colobus monkeys at Gombe National Park, and other less aggressive monkeys elsewhere.[37]

As described by Jane Goodall, chimpanzees dispatch their prey in rather gruesome ways as soon as they are captured:

> Chimpanzees kill their prey by (a) biting into the head or neck, (b) flailing the body so that the head is smashed against branches, rocks, or the ground, (c) disemboweling it, or (d) simply holding it and tearing off pieces of flesh (or entire limbs) until it dies. When several chimpanzees converge upon a small animal, it may be literally torn to pieces within moments of capture. Small prey, such as infant or juvenile colobus, black infant baboons, and striped piglets are usually killed by eating. Since the brains of such victims are almost always consumed first, death is very quick.[38]

Chimpanzees may hurl rocks at adult monkeys or bushpigs that attempt to protect their infants from being grabbed.[39] As far as we know, however, they do not use rocks to break open bones or cut meat, as early hominins did. In fact, contrary to chimpanzees' pervasive use of sticks and rocks for extractive foraging, they seldom use *any* kind of deliberately fashioned tool for hunting.

Seldom but not never.[40] Chimpanzees from the site of Fongoli, Senegal, provide an eye-opening exception. Unlike chimpanzees that live in forests, these savanna chimpanzees inhabit a mosaic savanna environment that is made up of small patches of gallery forest (i.e., bordering rivers), woodland, and grassland, which more closely resembles the habitats of early hominins[41] than those of modern forest-dwelling chimpanzees.[42] Like other chimpanzees, those from Fongoli hunt bushbuck fawns and a variety of monkeys without using tools, and males chase after monkeys more than females do. Only males at Fongoli hunt patas monkeys, while only females capture the terrestrial banded mongoose. There is one dramatic difference between the hunting styles of forest-dwelling and savanna chimpanzees, however. As famously discovered by anthropologists Jill Pruetz and Paco Bertolani,[43] only savanna chimpanzees sometimes use tools to hunt, and only to hunt for their most favored prey, small primates called lesser bush babies (*Galago senegalensis*).[44]

Lesser bush babies are small nocturnal prosimians (around seven ounces) that sleep during the day in hollow cavities in tree trunks or branches. Rather than being sedentary, these bush babies are highly mobile when awake and potentially aggressive, which is probably why Fongoli chimpanzees seem wary of being bitten by them.[45] After locating a cavity that might contain a bush baby, a hunting chimpanzee breaks off a living branch and removes its side branches and leaves. Sometimes the ends are further trimmed, and the bark stripped from the tool. The tip may also be bitten into a sharp point (figure 5.6). Chimpanzees use these spears by holding them in one hand and forcibly jabbing them into a hollow multiple times, after which the tool is removed and smelled or licked. In the first report of a successful bush baby hunt, after thrusting the spear into the cavity, the chimpanzee huntress reached in with her arm and removed the immobilized (possibly dead) prosimian.[46] A more recent report notes that, although male chimpanzees use tools to hunt bush babies, they also chase them down after they have been flushed out of their cavities by other tool-using individuals.[47] This suggests that these "spears" can also function as rousing tools. Remarkably, female savanna chimpanzees engage in most of the tool-assisted hunting of bush babies, which "permits individuals other than adult males to capture and retain control of prey."[48]

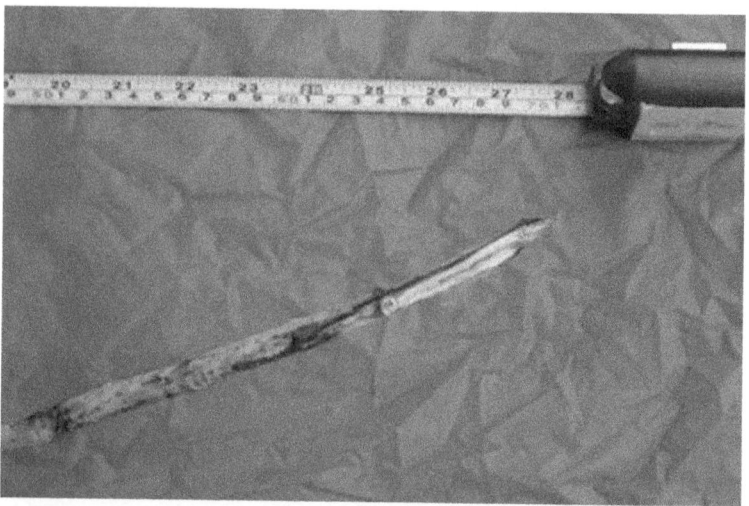

Figure 5.6. Tip of a two-foot-long chimpanzee hunting spear sharpened at the end with teeth. Reproduced from Pruetz and Bertolani 2007.

The differences between males and females at Fongoli are consistent with observations that wild forest-dwelling female chimpanzees appear to use termite-fishing poles and stone hammers more often and more skillfully than males.[49] Some scientists attribute this to their reluctance to compete with bigger and stronger males when it comes to the riskier (but toolless) business of chasing down, capturing, and killing fast-moving monkeys – and then keeping others from stealing the carcasses! Although female savanna chimpanzees apparently have less "brawn" for such pursuits, they seem to compensate for it, at least in part, by intelligently using botanical tools.

The Ugalla region of Tanzania is another rare seasonal savanna–woodland habitat that is home to non-forest-dwelling chimpanzees. Convincing evidence from three sites there shows that chimpanzees use tools in the form of sticks, wood, and bark to dig for a variety of tubers and roots (nutritious "underground storage organs"). Until this report, the use of digging sticks to acquire such foods was not known in apes. The chimpanzees' tools are flimsy compared to the larger wooden digging sticks of modern human foragers but are nonetheless able to penetrate the ground's hard surface. Chimpanzees probably enlarge the holes and remove the tuber or root with their hands, as indicated by the shallowness of the holes and the nearby finger drag marks.[50] Because the chimpanzees were not observed using the tools, it is not known if one sex did so more proficiently.

Sticks, Stones, and Early Hominins

Even though the archaeological record is silent with respect to the Botanic Age, paleoanthropologists are comfortable with the idea that the earliest hominins used sticks and stones to go about their business, much as living chimpanzees do. The fact that some savanna chimpanzees use digging sticks to obtain tubers and fashion wood spears for hunting bush babies suggests that these two types of botanic tools would have been important for early hominins as well, as they adapted to mosaic savanna–woodland habitats during the Late Miocene Cooling at the beginning of the Botanic Age (chapter 2).[51]

A relationship between these two types of tools is reflected in the observation that "a simple wooden hunting spear ... is but a lengthened digging stick."[52] Spears continued to develop during the Stone Age, as beautifully illustrated by eight long, carefully crafted and precisely balanced throwing spears made of spruce and pine and six double-pointed stick fragments that were discovered at a horse-hunting archaeological site in Schöningen, Germany (figure 5.7). This

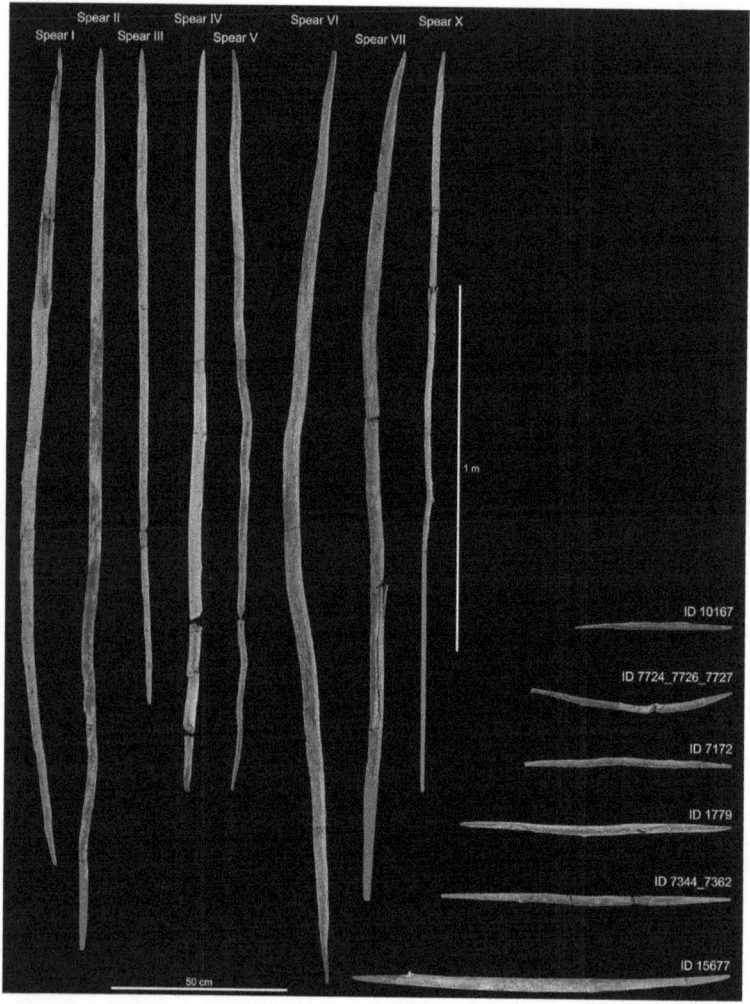

Figure 5.7. Eight spears and six double-pointed sticks from Schöningen, Germany. Reproduced from Leder et al. 2024.

site has yielded the largest assemblage of wooden artifacts to date and provides crucial understanding about the role of woodworking in human evolution.[53] The Schöningen spears are around 300,000 years old, while a tip from another spear from Clacton-on-Sea, England, may be 400,000 years old,[54] making it one of the earliest deliberately made wooden tools recovered from the archaeological record to date (figure 5.8).

The Clacton spear is not the oldest wood artifact in the record, however. That honor goes to a piece of polished plank from a willow tree in Israel dated to more than 780,000 years ago, although it's not clear what it was used for.[55] More recently, the earliest known hominin-crafted wooden structure dates to 476,000 years ago from Kalambo Falls in Zambia and consists of two logs that were deliberately notched to interlock.[56] Construction of such a structure would have required the use of stone hand axes. Given the location of the interlocked logs in a periodically wet floodplain, they might conceivably have been part of "a raised platform, walkway or foundation for dwellings."[57] This discovery, which was announced in October 2023, bowled over archaeologists because it not only added to our understanding of the technical cognition of hominins who lived nearly half a million years ago, it also highlighted "the role of this most humble of materials in the human story"[58] – namely, wood.

The widely accepted idea that the use of tools differed fundamentally between the sexes during hominin evolution is embodied by a saying handed down from Aboriginal women in Australia: "*We carry digging sticks, not spears. We are not men.*"[59] This notion may seem at odds with the fact that female savanna chimpanzees engage in most of the tool-assisted hunting of lesser bush babies. If chimpanzees in general, and savanna chimps in particular, provide reasonable models for identifying and assessing tool use in early hominins, then there

Figure 5.8. The Clacton spear, dated to about 400,000 years ago. By Geni, CC BY-SA 4.0.

appears to have been a role reversal. Females shifted away from being the predominant users of wood spears for hunting, and this eventually became the purview of hominin males. (Recall the photograph in the last chapter [figure 4.7] of a woman bearing a heavy load walking behind a man carrying a spear.)

This reversal is perhaps more apparent than real, however. Female chimpanzees use spears to obtain bush babies that are in holes – they extract them rather than chase after them. The oldest spears in the archaeological record, on the other hand (see figure 5.7), are thought to have been thrusting or throwing spears, or both, that were likely used to drive big game into trapped positions or ambush them.[60] Indeed, throwing spears were a potentially life-saving invention that allowed hunters to pursue and spear large animals from a safe distance rather than up close. Even if the Clacton spear was a thrusting spear, rather than throwing spear (as some believe), it "should not be seen as a less important weapon than a [throwing] javelin, since it is likely to imply cooperative hunting, not to mention extreme intrepidity."[61] This brings us full circle to the question of how a sexual division of labor for obtaining food evolved in early hominins.

Man the Hunter, Woman the Gatherer

It is important to keep in mind that chimpanzees obtain a vast majority of their food by eating fruits and leaves on the go, rather than from extractive foraging (mostly done by females) or hunting (mostly done by males), both of which are comparatively rare in apes. Human foragers, on the other hand, acquire most of their calories from collected and extracted foods, such as honey and tubers, and from regularly hunting a large variety of animals. Importantly, humans share food much more than chimpanzees. Similarly, though, there is a sexual division of labor in human foraging societies. Men are responsible for most of the hunting, especially for large and mobile game; women (often accompanied by children) gather small and relatively immobile food, which may include not only plants, but animals like piglets and other small mammals, nesting birds, tortoises, and so on. Nonetheless, women may help in men's hunts in various ways. Among the Hadza and Ju/'hoansi in Africa, for example, women sometimes track wounded animals and carry meat from the men's kills back to the community. However, these women are not known to hunt alone or with projectiles, or to kill large game.[62]

There are, of course, exceptions to the "woman-the-gatherer" generalization. For example, the Nanadukan Agta women on the island of Luzon in the Philippines frequently hunted for large animals in teams, using dogs to drive and immobilize them. This appears to reflect the community's recent practice of bartering meat for the botanicals they once collected.[63]

Some anthropologists emphasize the cultural rather than biological underpinnings of sex differences in human hunting. They note that, almost universally, men hunt with weapons like harpoons, spears, and arrows, which are formally prohibited for female hunters, who must rely on tools made for other purposes. If so, "it's not the hunt that is forbidden to women but, rather, the weapons."[64] This intriguing idea coincides with the fact that the meat acquired by men is often more valued in foraging societies than the botanical foods gathered by women. It has been suggested, reasonably I think, that if botanicals had been more valued than meat in foraging societies, the use of baskets and digging sticks would have been reserved for men![65] Obviously both nature and nurture were important during hominin evolution. What is remarkable is that the cumulative evidence strongly suggests there was continuity in the sexual division of labor from our earliest apelike ancestors through the emergence of *Homo sapiens*. Indeed, many would argue that men's work all too often continues to be more valued than women's, even in contemporary industrialized societies.

In any event, the comparative primate and ethnographic literature clearly indicates that hominins developed a taste for hunting much more than their apelike ancestors had, and that tools became increasingly important for hunting over time. The first tools designed for hunting were likely wooden spears; the use of stone tools for hunting emerged more recently. During the Botanic Age, females may have been using sticks and wooden spears to acquire food even more than males. Whether females invented the first spears is not known. However, given the high probability that females invented botanic baby slings (necessity is the mother of invention) and that female savanna chimpanzees hunt more with spears than males, we cannot rule it out.[66]

As we have seen, baby slings,[67] other carriers made of natural fibers, improved wooden spears, and the emergence of knapped stone tools were all important innovations during hominin evolution. As transformative as they were, however, they may have represented cognitive "hops" more than "leaps." The changes in prehistoric infants' physical development that prompted the invention of baby slings also affected how mothers and infants communicated with each other. This set the stage for the emergence, millions of years later, of perhaps the single most important factor in the evolution of human cognition – namely, language.

6

Babies Fall, Language Rises

SO FAR, WE HAVE CONSIDERED HOW botanical tools were likely invented and improved during the Botanic Age, with the earliest known stone tools as more recent additions to the hominin tool kit. We have also questioned the premise that stone tools were somehow superior to botanical ones, which is fraught with gendered assumptions about occupations that are often male-biased (alas, even today). If there was indeed a great "cognitive leap" that put our ancestors on the path to becoming humans, I believe it did not happen a mere 1.6 million years ago when hominins figured out how to knap fancier stone tools. To my mind, the most crucial leap occurred much earlier with the invention of a tool that, unlike baby slings or Acheulean hand axes, was intellectual rather than material.

The single most important thing that sets humans apart from all other animals is the fact that people use highly complex, symbolic, and grammatical language to think and express a potentially limitless number of ideas. Since apes never develop humanlike language, no matter how hard humans try to encourage them,[1] the ability to understand and express an infinite number of ideas by listening to and systematically producing sounds must have emerged at some unknown point after our predecessors diverged from the ancestors of chimpanzees. Thanks to Darwin, we know that the fundamental reason language evolved was because it was (and is) adaptive. In other words, prehistoric individuals who acquired language had a leg up when it came to matters that could make the difference between life and death. Those who survived passed along their traits – including ones that underpinned language – to future generations.

The invention of language would have served as both a stimulus for and a product of *Homo sapiens*' highly evolved brain.[2] But it did not happen overnight. I have long believed that the first spark for the eventual invention of language happened when baby hominins began to lose the ability to hang onto their mothers with their feet. Losing this ability threatened not only their lives but also those of their descendants – including us. In 2004, I proposed that vocal and gestural interactions between prehistoric mothers and their vulnerable infants initiated a prolonged sequence of events that eventually resulted in the emergence of baby talk (also called infant-directed speech, or "motherese") and, much later, the earliest simple language.[3] The reasoning behind my "putting the baby down" (PTBD) theory was that, because natural selection for upright walking inhibited the development of grasping feet, hominin babies were unable to cling to their mothers as all living infant apes and monkeys do. Therefore, ancestral hominin mothers must have actively carried their nurslings with them while they foraged and would have needed to put them down from time to time to gather resources and rest.[4]

According to this theory, hominin babies that were put down would have complained vigorously with gestures and cries (as modern babies are wont to do) and these, in turn, would have prompted mothers to respond with their own gestures and soothing vocalizations that, over time, "replaced cradling arms as a means for keeping the little ones content."[5] In this way, mothers' vocal utterances served as proxies for actual physical contact. This would have been a two-way street. Once reciprocal vocal and gestural utterances between ancestral mothers and infants got going, these kinds of communications evolved and eventually seeded the emergence of lullabies, simple baby talk and, further down the road, the very first language. Although the PTBD theory drew from wide-ranging studies of infant and child development, psychology, psycholinguistics, comparative anatomy, primatology, ethnology, and paleoanthropology, recent research shows that it needs some updating.

That's a Wrap: The First Slings Were Likely Botanical Straps

For starters, the PTBD theory failed to consider the role that baby slings must have played during the earliest and most mysterious part of hominin evolution – the Botanic Age. At the time, I believed that baby slings were not invented by the earliest hominins, but instead appeared much more recently in *Homo erectus*, around 1.6 million years ago.[6] Having made the mistake of thinking that absence of evidence for textiles in most of the archaeological record (chapter 5) was evidence for their absence, I thought that *Homo erectus'* baby slings were probably constructed from animal hides acquired from hunting or scavenging. Not only had the PTBD theory glossed over the time between 6.5 million and 3.5 million years ago (which I had not seriously considered until this book), it put the cart before the horse by implying that baby slings were invented *after* infants had lost the ability to cling: "Before baby slings were invented, ancestral mothers of helpless infants almost certainly would have put their babies down to dig for tubers, harvest berries, or just gather a handful of flowers."[7]

In light of new revelations, however, it seems likely that baby slings were invented as youngsters slowly lost their grips during the Botanic Age, rather than much more recently in *Homo erectus*. As discussed in chapter 4, the first portable tools invented by early hominins were probably simple baby slings made from vines or other pliable vegetal matter after our early African relatives began spending more time on the ground.[8] But this would not have happened quickly. By the time hominins began descending to the ground, they already understood the usefulness of botanical matter for counteracting gravity (sleeping nests). So it makes sense to speculate that mothers would have responded to their infants' increasing tendencies to fall by strapping them aboard with vines or other flexible botanical materials.

Ethnographic evidence shows that after *Homo erectus* began migrating out of Africa to regions with colder climates, baby slings became more complex and were constructed from a greater variety of natural materials, including wood and leather (chapter 4). Baby carriers of many hunting and gathering people were eventually designed for the dual purposes of carrying babies on mothers' backs or hips, on the one hand, and acting as blankets, bassinets, or cradles after being removed, on the other.[9] It seems reasonable to think that baby slings and carriers had similar functions in the past. After all, living great apes often make relatively simple day nests on the ground for brief rests[10] (chapter 2), a habit that was presumably shared by both their predecessors and early hominins. It would have been natural for baby slings to serve double duty then, as now, as carriers and also as little day nests that could be used to cushion infants when mothers put them down nearby on the ground. If so, baby slings were integral to mothers both carrying their infants and putting them down.

The PTBD theory also failed to consider the precise nature of the physical and behavioral changes that evolved in hominin babies during the Botanic Age, as they made a transition from clinging securely to their traveling mothers to being passively carried and put down on the ground from time to time.[11] This lapse was partly due to the fact that there were few, if any, relevant archaeological or paleoanthropological records from that mysterious time. Fortunately, new research has shed light on the evolution of babies that is useful for revising and extending the PTBD theory – which is how science is supposed to work.

Reflexes in Human Infants – Vestiges from the Past

Recall that in chapter 2 we explored the physical changes that accompanied the emergence of bipedalism by comparing milestones in chimpanzee and human infants, such as the ages when they are first able to hold their heads up, stand, take steps, and walk unassisted.[12] As we saw, ape infants reach these milestones much sooner than human babies, supporting the view that physical development in our ancestors' infants was delayed as their anatomy underwent adaptations for walking on two legs instead of four limbs. Put simply, when it came to the physical maturation needed to move in an adult manner, hominin youngsters were "late bloomers" compared to apes.[13]

If you have ever taken a baby for a medical checkup, you may have noticed the pediatrician eliciting certain motor reflexes to evaluate the development of the baby's nervous system. These reflexes are also interesting to evolutionary biologists, who compare them in ape infants and human babies to understand their different developmental pathways – which, as we will see, have evolutionary implications. For example, primatologists have observed that, unlike human babies, ape infants are able to sustain clinging to their mothers because "they come equipped with a different set of reflexes."[14] Bonobo chimpanzees are able to cling reflexively when they are born, which inhibits the hand-waving and foot kicking that are typical in human babies. In other words, bonobo infants never experience a stage where they lie on their backs and watch (and feel) their hands and feet moving above them.[15] This means that although our little ones may be enthralled by observing and wiggling their tootsies as they lie on their backs with their feet in the air, it's not so for tiny chimpanzees.

Although human babies cannot sustain unaided clinging like chimpanzee infants, this does not mean they completely lack an ability to grasp. If you place a finger in the palm of a tiny human's hand, you will probably elicit a (palmar) grasp reflex that is normally present during the first few months of life, after which it diminishes and disappears as voluntary use of the hands takes over (figure 6.1). Young babies also respond with a similar (plantar) grasp in which the toes automatically curl under when a thumb is pressed against the sole of the foot. This reflex fades around the time infants begin to stand unassisted. Grasp reflexes in both the hands and feet usually disappear by the time human babies reach their first birthdays.

Unsurprisingly, nonhuman primate infants have much stronger grasp reflexes than human babies, which contributes to their ability to remain slung beneath traveling mothers' bellies and, when they're a bit older, to ride piggyback by clinging to her hair with their hands and feet (see figures 4.1 and 4.2). Based on comparisons of these reflexes in humans and nonhuman primates, many neurologists think

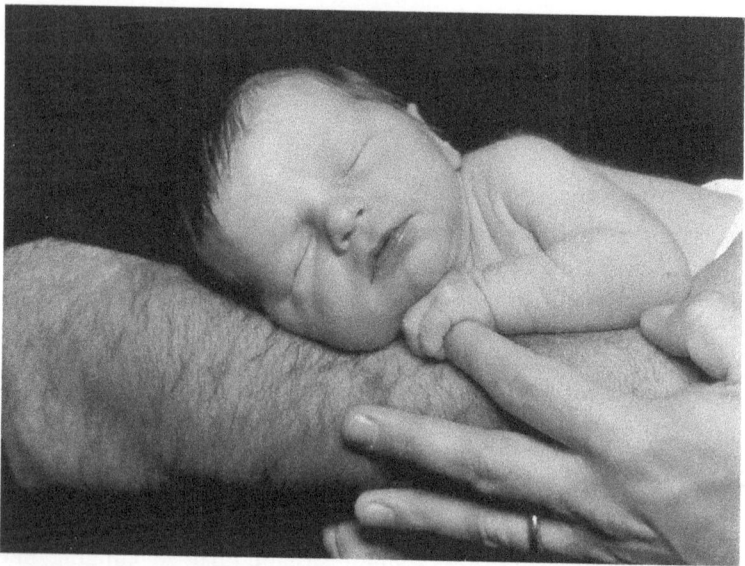

Figure 6.1. The palmar grasp reflex. Shortly after birth healthy babies automatically grasp a finger placed in the palm of their hand. This baby is the author's grandson, Jacob Riddle, when he was one day old. Reproduced from Falk 2009.

that the grasp reflexes in human babies' hands and feet have lost their usefulness and are simply vestigial retentions of reflexes that were once crucial for our early ancestors' survival in trees.[16] This makes quite a bit of sense.

Backward Hang Glider Landings

Sleeping human babies are likely to experience spontaneous, involuntary reflexes when their blood oxygen levels get too low – which can happen, for example, if they press their noses or mouths against bedding or other objects.[17] In response, the baby takes a deep breath and sighs, followed immediately by a startle reflex in which the head jerks and the limbs are thrown out (with fingers splayed) and then quickly brought back in. This is called a "sigh-startle" reflex.[18] Sigh-startles are often followed by thrashing of the limbs and head, after which the baby may cry, open its eyes, and awaken.[19] These linked reflexes freshen the baby's air supply.[20] Needless to say, maintaining enough oxygen in the blood while sleeping is crucial for babies' (or anyone's) health, so it is not surprising that sigh-startle reflexes are of keen interest to clinicians, including those who do research on sudden infant death syndrome. The sigh-startle reflex is typically diminished by the end of a baby's first year.

In an insightful analysis of the PTBD theory, David Lindsay interprets the sigh-startle reflex (and some other reflexes that eventually fade in developing humans) as a hand-me-down of reflexes that evolved in response to the gradually increased threat of falling that ancestral infants endured during the Botanic Age:

> Falk's [PTBD] hypothesis ... omits what I believe to be a significant piece of the puzzle ... Between the closest quadrupedal ancestor of humans and the first fully upright hominin lies a period of some millions of years. If hominin infants ceased to be able to cling to their mothers' underbellies as a result of bipedalism, the change is not likely to have occurred from 1 day to the next ... It would likely have been preceded by a phylogenetic phase during which infants held onto their mothers more tenuously. This would have been true regardless of the driving force behind bipedalism, due simply to the gradual nature of the progression from clinging to no longer being able to cling. *For hominin infants, in the glacial process of evolution, there must have been a protracted period marked by the anticipation of falling from their mother's underbellies, strategies for avoiding those falls and failures to avoid them*, with corresponding reflexive responses adapted to these conditions.[21] (emphasis mine)

Lindsay reasons that as hominin infants were becoming adapted for bipedal walking during the Botanic Age, they would have become heavier as they grew, and thus more likely to lose purchase on the underbellies of their traveling mothers. No doubt this would have been exasperating for both infants and mothers. Sighing, which is known to decrease physiological tension, would have relaxed infants' limbs and loosened their grips, resulting in a linked startle reflex that, as noted, ends with infants jerking their arms inwards (consistent with clinging). Recall that, like apes, attached hominin infants would have faced their mothers' chests, so the backs of their heads would have fallen backward when they fell. Allowing the head to drop back elicits a startle response in modern babies (indeed, this is one way doctors test for it), as does inhibiting the palmar grasp in newborn apes and monkeys.[22] Therefore, the startle reflex may be seen (among other things) as a response that counteracted babies' loss of grip during evolution.

Lindsay suggests that the thrashing reflex that follows a startle may also be a vestige of a competing urge to cling versus to break the fall. Furthermore, he suggests that the air-stepping reflex that two-to-six-month-old humans engage in when held aloft may have been evolutionarily associated with infants falling off traveling mothers. Experiments have shown that this reflex is strongest in babies that are suspended over treadmills with checkerboard patterns that appear to be moving away from them (consistent with what infants would see when falling backward off their mothers).[23] By considering all of this from an evo-devo perspective, Lindsay makes a compelling case that these various motor reflexes enabled babies to survive falling feet first off their roaming mothers' underbellies. As infants lost their pedal grips and began to fall backward, they would have automatically air stepped and (hopefully) caught up with their feet as they hit the ground. Lindsay picturesquely likens these maneuvers to "backward hang glider landings."[24]

Lindsay's research makes it clear that the PTBD theory did not give enough consideration to the physical evolution of babies that lived during the Botanic Age, which is what caused them to become separated from their traveling mothers in the first place. Indeed, it is easy to imagine a tiny Botanic Age hominin that was precariously attached to its traveling mother periodically losing its purchase, and ending up doing a backward hang glider (or other) landing. It would not be surprising if the unceremoniously detached infant complained vocally with whimpers and screams, just as the chimpanzee Bee-hinde did when she repeatedly fell from her polio-stricken mother (chapter 4).

It would have been natural for the mother to respond to her baby's protests by picking it up and making soothing vocalizations. (Considering this, perhaps the PTBD theory should be redubbed the "picking the baby up" theory!) According to the PTBD theory, it was these reciprocal emotional vocalizations between separated infants and their mothers that paved the way for the eventual invention of motherese, in which emotional meanings were not just expressed by tone of voice (affect), but also by words (linguistically). ("Oh, you poooor little thing, did Ba-by get a boo-boo?") Lindsay's evo-devo research shines a spotlight on mother–infant interactions that occurred much earlier in the hominin story than previously imagined – namely, at the beginning of the Botanic Age when bipedalism was emerging. As we will see, this shift in timing turns out to be extremely important.

Out of the Mouths of Babes

Mammalian mothers, including primates, are generally the primary caretakers of their young infants, so naturally this would have been the case for hominins during the Botanic Age. Because contact calls between separated infants and their mothers are common for mammals and even some birds,[25] it is a safe assumption that prehistoric infants who tumbled to the ground expressed their indignation vocally, and that their mothers responded in kind. Such reciprocal communications would have been good targets for natural selection because they would have been vital for survival then as they are now.

Human babies cry spontaneously when they are distressed or uncomfortable – when they are separated from caregivers, in pain, ill, or when they simply want something.[26] Because their cries are flexible, infants convey their moods and emotions (positive, negative, or neutral) by modulating their intonations, stresses, and rhythms – a component of speech known as prosody. Newborns also make non-crying sounds (such as coos, raspberries, growls, squeals, and isolated babbles) known as protophones,[27] which can be modulated. Although babies' vocalizations may not seem like music to parental ears, most of them contain very simple melodies[28] that develop into more complex ones over the first six months of life.[29] Eventually, these complex melodies contribute to the bits and pieces of sound that become the building blocks for babbling, in which babies utter consonant and vowel sounds (*ba, ga, da*), lengthened repeated sounds (*ma-ma-ma-ma-ma*) and, later, more complicated combinations (*e-goo, a-ba da-ba do*).

Babies' cries and other vocalizations, including their babbles, develop during intensely social interactions that entail imitation, turn-taking, and extensive use of melodic baby talk on the part of caretakers, which universally helps infants learn their languages.[30] In fact, these exchanges are why babies eventually (and adorably) babble in conversational tones of voice that use the melodies and rhythms of their native languages,[31] and how they learn to take turns "conversing."[32]

Although mature babbling sounds very much like speech, it is not, for the simple reason that babbles do not have specific meanings. Accomplished babblers are on the very brink of acquiring language. But to do so, they must realize that the sounds people make are symbolic and can be strung together. This "naming insight" frequently happens during a baby's second year in response to feedback from excited caretakers – for example, a mother who responds to her baby's repetition of the simple babble *ma* by enthusiastically pointing to herself and saying, "YES, *Mama*, that's me, that's me!" At this stage, babies begin to pick up new words by associating what they hear with what they see, for example when Daddy pats Fido on the head as he looks at Baby and emphatically says "*doggy*." Once the naming lightbulb goes on, there's no stopping Baby from learning and saying other words and, eventually, stringing them together into meaningful phrases and sentences. By then, Baby – who was born without language – is getting it.

Infant-directed speech typically sounds slower, higher pitched, and more melodic, and it has shorter sentences and longer pauses than adult-directed speech (ADS). Baby talk may be thought of as "simply... an exaggerated version of emotional ADS."[33] In other words, many differences between baby talk and adult talk may be thought of as a matter of degree rather than kind. There is, however, one feature that is unique to baby talk, according to Mitsuhiko Ota and his colleagues.[34] In languages across the world, baby talk has sets of unique words that are not usually used by adults when they talk to each other,[35] such as (in English) *doggy*, *choo-choo*, *tummy*, and *wee-wee*. Ota points out that the endings of diminutive words like *doggy*, *kitty*, *tummy*, and *daddy* sound the same, which provides cues about where words stop and start in the speech stream (segmentation). Many baby-talk words consist of replicated syllables, such as *din-din* and *night-night*, which also helps infants parse what they hear. These are easier sounds for them to remember than the adult versions of the same words. Ota reasons that baby-talk words help infants "overcome the initial difficulties in ... identifying the referents of word forms, and detecting word units in running speech."[36] Unique baby-talk words help babies make the transition from nonsensical babbles to meaningful words – from *choo* to *choo-choo* and (eventually) to *train*. Ota and his colleagues show that infants who hear more specialized baby-talk words develop larger vocabularies between the ages of 9 and 21 months.[37]

Baby talk is, above all else, fundamentally musical, which is related to the fact that it is highly emotional.[38] (This is why music is sometimes called the language of love.) In a sense prelinguistic infants are all about tone of voice – their own and their caregivers'. In the PTBD theory, the impetus for the (slow) evolution of language was when infants routinely started to be separated from their mothers. But it wasn't because mothers started to put them down nearby, as I once thought. Lindsay's evolutionary scenario not only places the separations of mothers and infants much earlier than I previously believed, but it also identifies changes in breathing as an important precursor for the emergence of babbling.[39] This seems apt because vocalizations in modern babies are all about breathing and control of the vocal cords. There are good reasons to believe that a variety of other anatomical changes also took root during the Botanic Age and that they eventually kindled language.

Bipedalism, Rhythm, and the Rise of Language

Human speech was facilitated by a larynx that descended lower in the neck during the evolution of bipedalism[40] and by evolved parts of the brain that control the vocal cords in the larynx.[41] These evolutionary changes are mirrored in the development of modern babies who usually do not speak until around the time they begin to walk, when their voice boxes have begun to shift lower in their necks (perhaps partly due to gravity)[42] and their brains have matured enough to exert (at least some) direct control over their vocal cords. Because the descent of the human larynx likely traces back to our ancestors' evolution of upright walking, it is safe to speculate that laryngeal changes in our lineage began in the Botanic Age when hominins were becoming bipedal.

Although humans and nonhuman primates use similar parts of their brains for triggering the involuntary reflex vocalizations that convey emotional information,[43] only humans have a second cortical region (called the dorsal larynx motor cortex [LMC]) that directly controls vibration and stretching of the vocal cords as air is exhaled or expired from the lungs.[44] Such voluntary vocalizations are necessary to produce meaningful speech (figure 6.2). Of course, much more happens during the development of babies' brains that paves the way for their acquisition of language during the first few years of their life, as detailed by many researchers.[45]

Many animals modulate their voices to express emotions and interact with other individuals (known as emotional and interactional prosody or EIP[46]) and the musical and rhythmic aspects of human speech evolved from similar, animal-like prosody. Linguistic prosody, however, has more layers than the EIP vocalizations of other animals, as detailed by Piera Filippi.[47] Animals "speak" largely through spontaneous emotional vocalizations; humans speak using voluntary symbolic vocalizations, which incorporate aspects of emotional prosody that make speech easier to produce and understand. (Tone of voice, it should be mentioned, is *loaded* with information about the speaker's state of mind that is potentially important for listeners.)

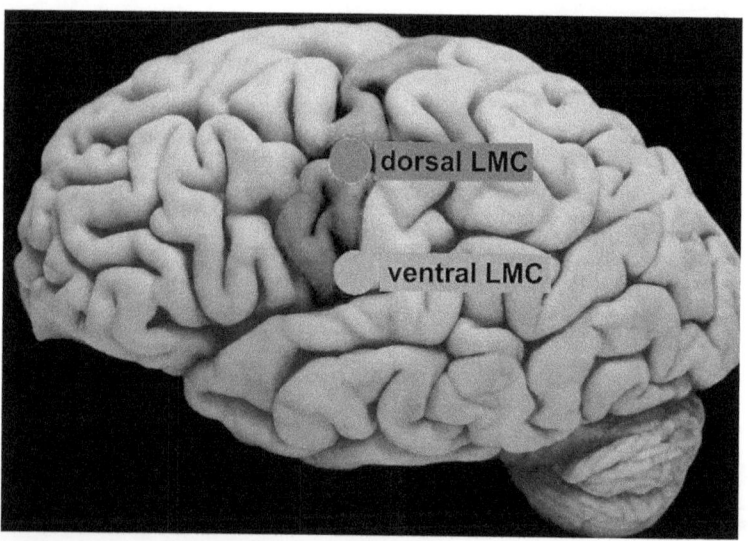

Figure 6.2. Left side of a human brain. Humans have two regions in the primary motor cortex (shaded) that control the larynx. The ventral larynx motor cortex (LMC) is related to emotional vocalizations and is shared with other primates; the dorsal LMC that controls the voluntary vocalizations necessary for speech appears to be novel in humans. Reproduced from Belyk and Brown 2017.

Rhythm that encompasses timing, accent, and groupings of sounds is particularly important for spoken language.[48] People speak in language-appropriate rhythms, are responsive to the rhythm of others' speech, and use rhythmic cues to take turns conversing. Compared to all other primates,[49] humans are good at perceiving a beat (e.g., in music) and moving to it (e.g., by tapping their foot),[50] an ability that was likely important for the emergence of the unique linguistic prosody that only humans have. For example, the ability of African great apes to voluntarily keep time to a beat appears to be quite limited, although they are known to drum rhythmically on trees or other objects and to produce some calls that have rhythmic elements (such as the breathy in-and-out panting vocalizations of chimpanzees known as pant hoots).

But how and when did humans get their sense of rhythm in the first place? A common assumption is that people develop rhythmic abilities while they are exposed to their mothers' heartbeats in the womb. This idea fails to account for *Homo sapiens*' unique rhythmic features, however, such as the ability to keep a beat. After all, human and ape fetuses hear similar sounds from maternal heartbeats. Swedish researcher Matz Larsson and his colleagues propose that the unique rhythmic sounds that human fetuses hear are the sounds of their mothers' footsteps. This contributes significantly to their development of musical and rhythmic behaviors, such as dancing, as well as a tendency to synchronize steps subconsciously when walking with others.[51] This makes sense because bipedal walking in humans produces rhythmic sounds that differ substantially from those made by quadrupedal moving in apes. Larsson therefore reasons that, from an evolutionary perspective, the transition to bipedalism was likely important for the emergence of advanced rhythmic abilities, because walking on two legs resulted in regular predictable stepping sounds. He further notes that hominin infants would not only hear (and feel) their mothers' rhythmic footsteps *in utero*, but also during the first few years of life as they were regularly carried in baby slings.[52] (More of Larsson's ideas are discussed in the interview with him at the end of the book.) However, it takes time for babies to fully hone their rhythmic skills. Human newborns perceive a beat, but do not move to it until they are around two years old.[53] By then, of course, babies are well on their way to acquiring language, thanks not only to rhythmic cues they absorbed while in the womb and bouncing along in baby slings, but also to the melodic baby talk they have listened to since birth.

After most of this book was written, Larsson and I began collaborating on these ideas, and how his research on keeping a beat and mine on motherese might fit together. We found that the evolution of feet was central to everything. First, feet modified for walking provided the literal basis for the upright rhythmic gait in gestating mothers that sparked hominins' evolution of beat perception. Second, the gradual loss of grasping ability in infants' feet was, again literally, responsible for their ultimate downfall (i.e., off their moms), which prompted the emergence of vocal exchanges that incorporated rhythmic beats and eventually seeded motherese. We therefore concluded that evolutionary changes in the nervous system and feet "contributed synergistically to the eventual emergence of rhythmically based behaviors like coordinated handclapping, chanting non-lexical vocables, and emergent motherese" and that these "subsequently underpinned the emergence of derived forms of music, dance, and language."[54] What this means is that, if our hypothesis is correct, language and music were minted as two sides of the same evolutionary coin. It also means that bipedalism was even more important for sculpting advanced cognition in our prehistoric ancestors than previously believed. Not to mention that it all began in the Botanic Age!

The threads that tie this package of evolved traits together are the little hominins who fell from their mothers and the mothers who responded in ways that kept their infants alive, kicking, and in the gene pool. One of the most important maternal strategies was to keep insecure babies attached to their bodies by using botanical materials (baskets on the hips); the other was to coinvent unique vocalizations with their infants, which may have served to ease their mutual unhappiness in the wake of the dirty trick Mother Nature was playing on them. There is reason to believe the invention of language, rather than sophisticated stone tools, was *the* major cognitive leap that set hominins on the path toward becoming the intelligent species that *Homo sapiens* is today. According to this view, mothers and infants were the inventors of the first language (protolanguage); it just happened a lot earlier than I (or anyone else) previously thought. An important implication of this is that the evolution of advanced cognition involved both sexes, not just (or mostly) males.

So far, we have envisioned how baskets in the trees (arboreal sleeping nests) became baskets on the ground (terrestrial sleeping nests), which eventually became baskets on mothers' hips (baby slings). As we have seen, it is highly likely that hominins first invented ground and hip baskets millions of years before they began to migrate out of Africa. Although hominins deliberately brought the practice of making and using these baskets to other parts of the world, there is another type of basketry that may have been important for colonizing far-off places. However, these botanical structures were not purposely transported abroad. Instead, during severe weather trees would have been uprooted and blown across bodies of water (as they are today). Some of these trees would have become floating rafts that harbored animal life. If these happened to contain sleeping nests with slumbering or sheltering hominins, they would have effectively been baskets in seas. If so, such baskets may be the key to solving one of the biggest mysteries in paleoanthropology – namely, how tiny hominins nicknamed hobbits came to colonize the Indonesian island of Flores.

7

What's Hobbit Got to Do with It?

IN 2004, PALEOANTHROPOLOGISTS WERE ASTOUNDED BY the discovery of a tiny new hominin species (*Homo floresiensis*) on the Indonesian island of Flores (figure 7.1).[1] The fragmentary remains of a handful of individuals who lived between 90,000 and 60,000 years ago[2] included a relatively complete skeleton of a 3.5-foot-tall woman, with the museum number LB1, nicknamed "Hobbit." Not only was LB1 much shorter than adults from the smallest living human populations in the world today, but her body also revealed a never before seen combination of features. Some characteristics resembled those of *Homo erectus*, who lived as recently as 112,000 years ago in nearby Java (figure 7.2),[3] while others resembled even older fossils of *Australopithecus* (or *Homo habilis*) from Africa.[4] Although she was bipedal, LB1 didn't walk like modern humans[5] and she likely spent time in trees,[6] including sleeping nests. No wonder scientists still argue about whether *Homo floresiensis*' ancestors descended from australopithecines, *Homo habilis*, *Homo erectus*, or another species.[7]

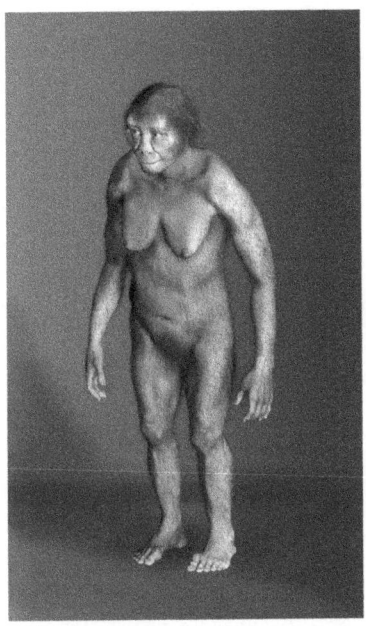

Figure 7.1. Paleoartist's reconstruction of LB1 ("Hobbit"), the most complete specimen of *Homo floresiensis*. Courtesy of Katrina Kenny.

When the discovery of *Homo floresiensis* was announced in 2004, a small group of paleoanthropologists argued that LB1 and the others in her group represented diseased *Homo sapiens* rather than a previously unrecognized species. The disease attributed to *Homo floresiensis* changed, however, as each new claim was refuted. Suggested pathologies included microcephaly, various forms of dwarfism, cretinism, and Down syndrome, none of which withstood scientific scrutiny.[8] As paleontologist Michael Westaway concluded, "Many interesting questions about the … fossils remain unanswered, but whether LB1 is a pathological *Homo sapiens* is not one of them."[9]

The discovery of *Homo floresiensis* rocked the paleoanthropological world because LB1 looked completely out of place in Indonesia, and the date seemed much too recent for such a tiny[10] and (in many

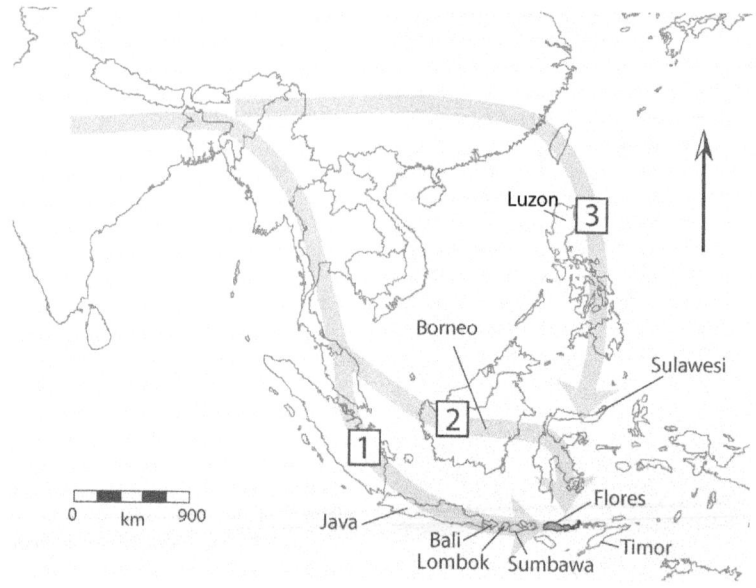

Figure 7.2. Possible routes of *Homo floresiensis* to Flores, Indonesia. Although it is not known exactly how hobbits got to Flores, scientists suggest three possible routes: (1) eastward from Java, (2) from Borneo via Sulawesi, and (3) from China via the Philippines and Sulawesi. Adapted from Dennell et al. 2014.

ways) primitive-looking species. Had LB1's remains been recovered in deposits that were a couple million years older in Africa, or if anything resembling it had previously been found in Southeast Asia, the discovery of *Homo floresiensis* might not have created such an uproar.

Because LB1 seemed so strange, the announcement of the species *Homo floresiensis* was highly controversial.[11] The discovery 15 years later of another startling tiny hominin that lived about the same time as LB1, but this time on the island of Luzon in the northern Philippines (see figure 7.2),[12] did little to abate the controversy.[13]

Named *Homo luzonensis*, it consisted of fragmentary teeth and limb bones from at least three individuals. Like their contemporaries on Flores, the new species had a unique mixture of traits not seen in any other hominin. Importantly, *Homo luzonensis*' australopithecine-like hands and feet suggest that, similar to *Homo floresiensis*, it may have spent time climbing, moving, and sleeping in trees.[14]

No one knows who *Homo luzonensis*' or *Homo floresiensis*' ancestors were or exactly where they came from. But although the two tiny species are not considered to be close relatives, their stories are similar. The predecessors of *Homo luzonensis* got to Luzon by at least 700,000 years ago when the first known stone tools appeared on the island, and the first hominins to arrive on Flores did so before the date of the oldest tools there, over one million years ago.[15] Similar stone tools were found on Flores near fragmentary remains of hominins dated to 700,000 years ago,[16] and more recently still, with LB1 and her kind. Some believe the earlier remains represent direct ancestors of *Homo floresiensis*, while others think the latter originated from a separate colonization.[17]

As with *Homo luzonensis* to the north, the colonizers of Flores had to cross large bodies of water to get there, although it's not clear which ones. Did LB1's predecessors come from *Homo erectus*' stomping grounds in Java, as some believe?[18] Or did they migrate from South Asia, through Indochina, Sulawesi, and then south to Flores,[19] swept along by the strong Indonesian throughflow ocean currents?[20] As the map in figure 7.2 suggests, it is conceivable that, in addition to *Homo floresiensis*, the hominins that colonized Java (*Homo erectus*) and the Philippines (*Homo luzonensis*) may have also started out in Asia.

Whatever bodies of water the founding hominins crossed to get to Luzon and Flores, many believe colonization of the islands was the result of events such as tsunamis. Perhaps a small colonizing group accidentally arrived on the islands after clinging to a natural raft of matted vegetation, or occupying a sleeping nest in an uprooted tree that had been blown out to sea.[21] As one colleague colorfully imagines, Flores' first arrivals could have included "a pregnant *Homo erectus* [who was] washed downstream and out to sea while clutching organic detritus during the monsoon."[22] In any event, given that the earliest hominins arrived on Flores more than one million years ago, most scholars think that the predecessors of *Homo floresiensis* (and presumably *Homo luzonensis*) arrived on the island "by accidental drifting rather than from purposeful navigation."[23] Deliberate boat building would come much later.[24]

Indeed, accidental "rafting" is thought to have contributed to the colonization of islands in Indonesia[25] and elsewhere by many animals – including some primates, such as the ancestors of New World monkeys that arrived in South America from Africa around 40 million years ago.[26] It is well known that natural disasters like tsunamis and tropical cyclones occur frequently in Southeast Asia and that they facilitate the rafting of various small animals. Of course, there would have needed to be enough original colonizers on Flores (and Luzon) to establish a lasting population, which has been defined as 500 individuals or a population that survives 500 years.[27] Computer simulations have shown that, given the length of time involved, less than 20 accidental hominin rafters could have conceivably established *Homo floresiensis* on Flores, especially if additional arrivals occasionally contributed new genes.[28]

Baskets in the Seas

But still, 20 hominins being blown out to sea by a tsunami or tropical storm and then beaching on the same island? How could this have happened? For starters, consider the aftermath of a tsunami in 2004 that washed over the Nicobar Islands that extend from the Andaman Islands to Sumatra. Three of the islands are home to long-tailed macaque monkeys (*Macaca fasicularis umbrosa*), which forage and lodge in trees next to water. A survey of 40 groups of monkeys showed that the number of groups decreased sharply after the tsunami, particularly those in coastal forests rather than interior areas, leading the study's authors to suggest that the main cause for the decline was likely the disappearance of several kinds of coastal palm and other trees where the monkeys liked to feed.[29] Palm trees are especially frequent vehicles for accidental rafting, at least for mollusks and other invertebrates in Indonesia.[30] Palms are particularly susceptible to storms and typhoons because strong winds tear them up by their roots[31] and, once they are swept to sea, they float upright so that their upper parts are not submerged in sea water.[32] Floating palm trees sometimes land near each other on beaches (figure 7.3), possibly because floating objects tend to be accumulated by ocean currents.[33]

Some of the missing coastal monkeys in the Nicobar Islands likely died during the tsunami, while others may have been swept out to sea in uprooted trees where they survived a little longer. Even if none of the monkeys from the 2004 disaster made it to other islands, this example illustrates how numerous species of macaques may have dispersed to the various islands they now occupy in Southeast Asia.[34] This explanation is plausible because it is common for uprooted trees from coastal forests – for instance, in the mangrove swamps and forests of Indonesia – to be blown into nearby rivers and carried out to sea during storms.[35]

The main features of tropical cyclones are heavy rainfall and high winds, and defoliation is their most common effect.[36] Cyclones can occur night or day. Although there are almost no data on how

Figure 7.3. Nipa palm rafts that landed on a beach on the northwest coast of Borneo. Reproduced from Raven 2019.

free-ranging arboreal mammals – let alone wild nonhuman primates – behave during tropical storms,[37] we know that great apes are likely to be in sleeping nests during storms that make landfall at night. Chimpanzees have also been reported to spend more time resting in trees during rainy seasons,[38] and observations at the Atlanta Zoo suggest orangutans seek shelter in nests on stormy days.[39] Further, great apes construct their tree nests in ways that protect them from inclement weather.[40]

Based on this information, it is reasonable to theorize that ancient hominins might have been transported unintentionally, but successfully, to Flores and Luzon in storm-tossed trees that contained their sleeping nests. This could have occurred during the Ice Age, when water became locked up in glaciers and water barriers decreased. We know that the climate in Indonesia and Southeast Asia was (and still is) marked by violent storms, cyclones, and tsunamis. Observations of the only great ape that lives in Asia today (the orangutan) suggest they take shelter in tree nests during stormy weather.[41] The trees they build nests in include types that are known to be uprooted and blown into rivers and oceans by storms and are susceptible to becoming unintentional rafts for their occupants.[42]

Anatomical similarities between hobbits, their cousins in Luzon, and today's great apes indicate that, like the great apes, the early hominins on Flores and Luzon likely slept in tree nests.[43] The African great apes make their night nests in neighboring trees. If ancestors of *Homo floresiensis and Homo luzonensis* did the same, it would have increased the likelihood that numerous individuals would have been impacted by the same cyclone or tsunami. This scenario accounts for how, given a great deal of time and repeated rafting events, enough individuals to found a population (20 or so) might have arrived on the islands. We know, of course, that many small species of animals have been transported to new habitats on fallen trees and other seaborne detritus. So why not these diminutive hominins?[44]

But how can scientists explore this idea that sleeping nests lodged in uprooted trees could have provided serendipitous vehicles for the hominins that populated Southeast Asia during the Pleistocene? More information about how great apes behave during severe rainy weather and tsunamis could be collected using strategically and securely placed cameras in zoos, fieldwork sites, and orangutan rehabilitation centers. Experiments could be conducted in Southeast Asia to delve into the question of what happens during cyclones to proxy sleeping nests (if not actual ones that have been abandoned) containing biodegradable surrogates for early hominins. In addition to careful planning, such research would require the use of sensors and a way to track them.

Fortunately, such research is already being conducted by experimental archaeologist Glenn Marshall who belongs to a group called the First Mariners that includes sailors, maritime historians, and early human migration researchers. The group builds watercraft from natural materials like bamboo, similar to those that humans might have used to colonize Australia around 65,000 years ago, and then tests them. Marshall's main role is to build and launch current drifters to gather information about the strong currents and winds and to monitor their impacts on voyaging rafts. He is currently applying his findings to the question of how *Homo floresiensis* ended up on the island of Flores. (You can read about this in an interview with him at the end of the book.)

Meanwhile, one thing is for sure. Early hominins had to cross bodies of water to get to Flores and Luzon, and it was not on synthetic materials like plastic and polystyrene foam that, unfortunately, form a significant portion of marine debris today.[45] But cross they did – and it is almost certain that the accidental rafts that ferried them consisted of botanic material. As we have seen in earlier chapters, ape sleeping nests resemble large baskets in trees. If the scenario here is correct, then what were once baskets in trees became baskets in seas during natural disasters in Indonesia, and a tiny fraction of them dispersed their occupants to distant lands. If so, the Botanic Age had profound, if inadvertent, implications for humanity's eventual colonization of far-off islands.

From Accidental Rafts to Deliberate Seacraft

Unlike the intuitive grasp of folk physics that prompted prehistoric mothers to invent baby slings, early hominins' use of tree nests as life

Figure 7.4. Dated at somewhat less than 10,000 years old, this dugout canoe from the village of Pesse in the Netherlands is, so far, the world's oldest known boat. The canoe, which is almost 10 feet long, is located in the Drents Museum in Assen, Netherlands. By Christoph Braun, CC BY 3.0.

rafts would not have been deliberate. Nonetheless, these diminutive beings likely had enough experience with inclement weather to take shelter in tree nests (assuming they weren't already asleep in them) and perhaps even to batten down the hatches (so to speak) during the violent storms and cyclones that plagued Southeast Asia then, as now. Nobody knows when hominins first came to the realization that wood floats and began making simple water vehicles on purpose. But once they were invented, numerous innovations in woodworking technology contributed to subsequent improvements of watercraft, from dugout canoes to the first known ships (figure 7.4).[46]

As described by Roland Ennos, farmers who lived during the New Stone Age (Neolithic) that began around 12,000 years ago developed a method of making wickerwork by "weaving a series of narrow shoots in and out at right angles through a parallel framework of thicker ones," which was used for many purposes, including the construction of "lightweight rounded boats with wickerwork frames and a leather hull."[47] According to Ennos, farmers may have used such boats to migrate across Europe between 10,000 and 8,000 years ago. Unlike floating tree nests, these "baskets" in the water were, of course, deliberate inventions.

The invention of watercraft and seafaring changed the world in dramatic ways. The first small boats were likely used to transport people and goods relatively short distances along rivers and across lakes. The eventual invention of much larger boats and ships permitted humans to ferry livestock and transport cargoes of several tons over great distances. Long-distance trade opened up, as did the potential for discovering and colonizing new lands. For example, plank ships allowed Mediterranean sailors to carry supplies across vast maritime regions.[48] The same was true for other regions around the Far East, India, and Arabia. Wooden ships contributed enormously to the growth of trade-based empires.

However, hominins may have used relatively simple seacraft constructed from botanic materials to colonize Australia well before the invention of wickerwork boats or the earliest known ships. Although archaeologists are not certain about the precise water route(s) through the Indonesian islands that *Homo sapiens* used to get to Australia some 65,000 years ago,[49] it is generally thought that they did so in deliberately built watercraft, which would have been much older than the current archaeological record for preserved boats.[50] That, of course, did not happen until the lightbulb went on and *Homo sapiens* realized that wood could be fashioned into floating vessels.

The vast regions of the Americas, on the other hand, did not have people until well after Australia was colonized, although archaeologists disagree about exactly when and how the first people got there.[51] Until recently, most scientists thought the first Native Americans traveled on foot from Asia over the Bering Land Bridge and down into North America around 15,000 years ago, began spreading rapidly across North America by 13,500 years ago, and then migrated down into Central and South America. However, recent evidence from footprints discovered in New Mexico suggests that people had arrived in North America by at least 23,000 to 21,000 years ago.[52] It is reasonable to assume that these migrants used wood to make fire and various botanical tools including weapons, baby cradles, and other carriers.[53] Another theory (not necessarily mutually exclusive) is that some colonizers arrived at an earlier time from northeast Asia by traveling along the west coast of the Americas, presumably in "relatively sophisticated watercraft."[54] This coastal route, once thought to have been an unlikely path for entering the Americas, is now gaining serious consideration as one of the earliest routes for doing so, perhaps as much as 25,000 years ago.[55] If confirmed, it would mean that wooden boats enabled humans to spread not just to Australia, but also to other huge landmasses.

As we have seen, although the Botanic Age ended around 3.5 million years ago, it extended its branches throughout its better-known successor, the Stone Age – both on land and at sea. It bequeathed a stunning legacy to our predecessors, from stationary hominin tree nests in Africa to portable baby slings made of vines to beached trees containing hominins stranded on tropical islands. After that, seeds of the Botanic Age sprouted even further into the Stone Age, leading to the first deliberately built rafts and, eventually, the invention of wooden ships that led to much of the world being colonized.

Beyond Boats and Botany

This book has focused on a number of life-altering botanic inventions that appeared during and after the Botanic Age, including baby slings, wooden weapons, shelters, wheels, canoes, and ships (chapter 5).[56] Other materials appeared more recently, including metals such as copper, bronze, and iron, which have their own ages named for them following the last part of the Stone Age. Although these materials impacted civilizations in important ways, they were outshone by one recent botanic invention that did nothing less than transform humanity's ability to process, retain, and pass knowledge along to future generations – an invention that literally made language visible. Paper.

As far as we know, reading and writing were invented around 5,500 years ago when bookkeepers began recording trades by inscribing marks on clay tablets in Mesopotamia (present-day Iraq).[57] This then spread rapidly to other parts of the world, where writing was scratched, engraved, embossed, or painted on stone, bone, metal, shell, or clay.[58] Before paper made from plants and shrubs was invented in China around 2,000 years ago,[59] paper-like scrolls of papyrus, made from the stalks of grass-like plants, were favored as a writing material, particularly in Egypt.[60] Because paper was easier to make, cheap, relatively permanent, and eventually lent itself well to the printing press, it more or less replaced other writing materials and "Europeans ... carried paper and papermaking, along with printing, throughout the globe."[61]

With the inventions of reading, writing, and excellent material to write on, people no longer had to depend on human memory and word-of-mouth to recall, archive, and distribute information. Paper had become the preferred medium for memorializing the potentially limitless number of ideas that only a linguistic species like ours can formulate, at least until quite recently.[62] In other words, paper, whose cellulose roots harken back to the Botanic Age, radically transformed the cognitive future of humankind. Within the last half century, humans have skyrocketed into the Digital Age, which would not have been possible without the initial invention of a material that made the retention and transmission of knowledge easy and widely available. When it came to inventing such a material, paper beat not just rock – but also everything else.

Conclusion

THIS EXPLORATION OF HOMININS' USE OF botanical tools has covered a good deal of territory and a vast amount of time. Its starting point was literally at "ground zero," when our earliest hominin relatives still spent most of their time in trees and wove arboreal sleeping nests from branches, leaves, vines, and twigs. As comparative evidence from nonhuman primates shows, this habit was underpinned not only by excellent weaving skills, but also by an unconscious sense that sleeping platforms prevented falling (intuitive "folk physics"). It is a good bet that tree nests provided the basis for the other botanic inventions, like baby slings, that arose during the first half of hominins' 6.5 million years of existence, and that these inventions had an enormous, if unintended, impact on cognitive evolution in the process.

Botanic inventions clearly preceded the first deliberately made stone tools by millions of years. However, the Botanic Age did not come to a halt after our ancestors figured out how to manufacture stone tools around 3.5 million years ago. The handful of botanic artifacts that have, against all odds, been recovered in the archaeological record show that the Botanic Age continued to exert influence throughout the Stone Age. As we have seen, the oldest known examples of deliberately modified wood are a 780,000-year-old polished plank from a willow tree in Israel and two deliberately notched and interlocking logs dated to 476,000 years ago in Zambia. Wooden spears have been preserved from as long ago as 400,000 years, while botanical clothing, wooden canoes, and wheels don't appear in the record until much more recently (around 5,000 to 6,000 years ago). Although these artifacts are much more recent than the earliest known stone tools, which are dated to 3.3 million years ago in Kenya, this is still an impressive record given that plant matter usually disappears over time (whereas rocks, of course, do not).

The traditional opinion has been that advanced cognition got underway long after the Botanic Age transitioned into the Stone Age. In fact, numerous cognitive archaeologists do not see signs of a "cognitive leap" in human ancestors until around 1.6 million years ago when hominins were shaping sophisticated (Acheulean) hand axes out of rocks. A subtle, and sometimes not so subtle, assumption has been that these tools were made mostly, if not entirely, by males.[1] While the sculpted Acheulean tools are indeed impressive, it is likely that the ability to knap them was just one of many products of the ongoing cognitive evolution that had begun considerably earlier in conjunction with botanic inventions. (This is not to say that males were any less important for hominin evolution than females – after all, when it comes to passing genes into the future, it takes two to tango.)

If this alternate view about the evolution of advanced cognition is correct, it was a whole new neurodevelopmental ball game for our ancestors after mothers and infants invented language.[2] With the ability to express an infinite variety of ideas, conscious symbolic thought took off big time, and other advanced forms of cognition like mathematics, scientific thinking, music composition, and reading eventually came into existence on the backs of evolved language networks and bilaterally reorganized brains.

Some would note that the utilitarian importance of botanical tools has been surpassed as other materials like metals have gained favor and new technologies have been invented – many of the most spectacular ones emerging around the mid-twentieth century. As with the Botanic Age and the Stone Age, the names of modern ages reflect these innovations. The Nuclear Age began with the first detonation of an atomic bomb in 1945, the Space Age was ushered in by the launch of Sputnik 1 in 1957, and the current Digital Age (also called the Information Age or the Electronic Age) was arguably up and running by 1980. In the face of such developments, it is not surprising that botanic inventions (including paper) have taken a back seat to innovations that were previously inconceivable.[3]

Nonetheless, this book has emphasized the invention of botanic artifacts that served specific purposes such as transporting objects and people from point A to point B, providing materials for constructing shelters and clothing, improving weapons, and creating a medium for memorializing and passing on knowledge. Moreover, a case has been made that some innovations, such as the invention of baby slings and the shift to full-time ground sleeping, contributed to the remarkable intellectual evolution of our species. However, there has been little if any discussion about the impact that botanic materials may have had on the less cerebral, more artistic, and spiritual evolution of hominins. It therefore seems fitting to end with a brief consideration of the legacy the Botanic Age may have left to these aspects of the human psyche.

Veneration of Botanic Matter

The connection between humans and plants is reflected in a deeply rooted reverence of trees and botanical inventions that appears in symbols and allegories throughout folklore, myths, fairy tales, nursery rhymes, fables, lullabies, and religious stories that continue to be passed from generation to generation.[4] For instance, most of us are familiar with narratives, such as the Tree of Knowledge in Christianity or the Bodhi tree in Buddhism, that are based on the belief that human knowledge and enlightenment came from trees.[5] The same can be said for stories based on the idea that wooden watercraft, such as Noah's Ark, were crucial for the survival and spread of humanity.[6]

At a more aesthetic level, environmentalists and others value the tranquility and beauty associated with spending time in botanical habitats, such as the Gurukula Botanical Sanctuary that conserves rainforest plants in Kerala, India; the old growth redwoods at the Muir Woods National Monument near San Francisco, California; and Wakulla Springs State Park, a 6,000-acre wildlife sanctuary not far from Tallahassee, Florida, that is famous for its riverboat tours that reveal not just pristine flora, but also the world's largest and deepest freshwater springs and stunning animal life.[7] The German language even has a specific word – waldeinsamkeit – for the sublime or spiritual feeling of being alone in the woods.

Some environmentally minded artists favor botanical matter for creating aesthetically pleasing and thought-provoking pieces. Danish artist Thomas Dambo makes amazing sculptures from recycled wood – namely, huge trolls that are installed in various outdoor settings around the world (figure 8.1). Dambo, who has made over 80 trolls (and counting), suggests wood is the best medium in the world for building things: "It just comes right out of the ground. It doesn't

Figure 8.1. Oscar the Bird King, an enormous sculpture created by Thomas Dambo from local recycled wood and branches on Vashon Island in Washington State. By Travelcompostions, CC BY 4.0.

pollute anything, it's not toxic to work with, and when it decays it provides more nutrient."[8] Dambo uses volunteers to construct his whimsical giants, which he deliberately puts in places that will motivate would-be viewers to venture into nature.

Other environmentally inclined artists rely more on the ancient art of weaving when crafting their botanical artwork. Patrick Dougherty and his son Sam constructed an enormous stickwork sculpture titled "Grand Central" mostly by using their hands to bend, interweave, and fasten together willow saplings that were harvested from a farm in New York State without killing any trees. The sculpture was installed in a large palm grove at McKee Botanical Garden in Vero Beach, Florida (figure 8.2). It's gone now, but the artists don't mind

Figure 8.2. A portion of the stickwork sculpture woven from willow saplings by artists Patrick Dougherty and his son Sam. Photo by Dean Falk.

because they expect the natural materials in their sculptures to break down eventually and become part of the landscape. To them, there is beauty in the ephemeral nature of botanic art.[9]

Weaving branches and twigs into baskets may well have been the most ancient craft invented by early hominins. Weaving seems to have left its imprint (quite literally, in some cases) on some very old objects in the archaeological record. Fiber expert Helen Anderson draws attention to impressions on shells and pottery of crosshatched, parallel, or zig-zag patterns that resemble the simple geometric patterns in weaving. These patterns appear incised on ironstone slabs at Wonderwerk Cave in South Africa and engraved on a mussel shell far away in Java around half a million years ago.[10] The appearance of these widespread geometric patterns supports the idea that basketry and weaving were the first crafts invented millions of years ago and have been handed down through the ages. If so, one might think of basket weaving as a prerequisite that provided the basics for inventing the other botanic tools that followed – a "Basket Weaving 101" course for early hominins that lived during the Botanic Age.

Unfortunately, "Basket Weaving 101" has long been used as an idiom for college or university courses that are easy to pass – so-called gut courses. This less than favorable regard for basket weaving is reflected in academia at large. Most academics pay little attention to how the arboreal lives of our earliest ancestors may have had a major influence on hominin cognitive evolution. Instead, when it comes to cognitive evolution, the standard textbook emphasis has been rather macho – on the (presumably) masculine production of stone tools for hunting and protection. Interestingly, the importance of botanic materials for shaping humanity seems to be more widely accepted in popular and religious narratives, as well as the creations of environmentalists and artists, than in the academic output of many professionals who study hominin evolution!

The staying power of the materials and weaving skills that were first used during the Botanic Age can be seen in the work of Pueblo fiber artist Louie García (figure 8.3). As you can see in his interview below, García's art is filled with historical and prehistorical significance: "As I sit at my loom, I am often singing a song in prayer as I weave good thoughts and good intentions into each weaving that will be used by an individual or for use in a ceremony to bring good things to the people." No one better captures the idea that "Basket Weaving 101" may have been one of hominins' most important courses, not just because it provided a foundation for cognitive evolution, but also because it sparked the eventual emergence of humanity's unique artistic and spiritual awareness.

Conclusion

Figure 8.3. Pueblo fiber artist Louie García at the loom. Courtesy of Louie García.

People behind the Book

A book that focuses on the first three million years of hominin evolution is bound to contain a good deal of speculation. After all, the fossil and archaeological record for that long period of time is essentially nonexistent. Readers might, therefore, be tempted to consider some ideas in *The Botanic Age* as "just so" stories, akin to Rudyard Kipling's fanciful bedtime tales ("How the Leopard Got His Spots," "How the Camel Got His Hump," and so on; see figure 9.1). However, credible books about the natural world must be rooted in evidence that supports their claims, which is a very different kettle of fish (to use a phrase that Kipling would have appreciated).

In keeping with this more scientific approach, this book owes a huge debt to natural scientists, social scientists, evolutionary anthropologists, ethnologists, comparative psychologists, linguists, experimental archaeologists, cognitive archaeologists, neuroscientists, paleoanthropologists, evolutionary developmental biologists, ethnomusicologists, artists, primatologists, and others. If you are interested

Figure 9.1. Cover of Rudyard Kipling's *Just So Stories*, 1912 edition. Copy held at the New York Public Library, scanned by Nicole Deyo.

in learning more about the kinds of evidence that support this book's argument for the identification of a new age, extensive information is provided in the notes for each chapter. If you are content not to delve into the scholarship behind the book, no problem, but you should know that *The Botanic Age* is not a "just so" story made up from scratch. (Unlike Kipling, alas, I do not have a vivid enough imagination for that.)

Instead, the theories presented here about the emergence and evolution of our earliest ancestors are firmly grounded in the research of some very smart people, eight of whom are interviewed below. I have asked each person three questions about their work, focusing on aspects that were important for developing *The Botanic Age*. The interviews are in approximate chronological order, from research on the earliest hominins to living humans, and each interviewee answers the questions in their own words. I believe these short exchanges not only illustrate the diversity of rigorous research in this field, but they also convey the joy and satisfaction of doing science and making art.

The Interviews

1. Matz Larsson	Maternal Footfalls
2. Kathelijne Koops	The Night Shift
3. Susanne Shultz	Nocturnal Predators – The Downside of Sleeping on the Ground
4. Cara Wall-Scheffler	Women – The Burden-Bearing Sex
5. Helen Anderson	Crosshatching, Trellises, and Diamond Patterns
6. Rebecca Biermann Gürbüz	Woodworking and Cognition

7. Glenn Marshall	Rafting to Australia and Flores – How'd They Do It?
8. Louie García	The Fiber Arts, Then and Now

1. Matz Larsson

Maternal Footfalls

Dr. Matz Larsson (figure 9.2) is an associate professor in the Clinical Health Promotion Centre at Lund University in Sweden. He has researched and published papers on the evolution of various traits such as vocal learning, right-handedness, and eye–hand control. Below, he discusses his research on the major role that bipedalism played in the evolution of the unique rhythmic abilities in hominins, which led to the universal emergence of dance, music, and speech.

After this interview was completed, Larsson and I began collaborating on a project about the earliest hominins that combined his research on the impact of maternal footfalls with mine on the emergence of motherese. We are hoping that the sum is more than its parts, as discussed in a paper that will appear in Current Anthropology *along with commentaries from international colleagues (Larsson and Falk 2025).*

> **FALK: You have theorized that the transition to a bipedal gait in early hominins may have been seminal for the evolution of unique human rhythmic capabilities such as dance, music, and the ability to keep a beat. How would this have transpired?**
>
> LARSSON: According to the acoustical advantage hypothesis, people with roughly the same leg length tend to move

Figure 9.2. Dr. Matz Larsson with his grandson. Photo courtesy of Matz Larsson.

unconsciously in tandem, which produces a distinct beat. One result of paced walking is the creation of short intervals between footfalls with relatively low noise levels during which one can more easily perceive sounds from the environment. Moreover, footsteps of a pair walking in regular pace are simultaneous and predictable. Although the footsteps come from two individuals, one perceives them as coming from a single source, which helps the brain distinguish footsteps from other sounds in nature. (The scientific word for that is *auditory grouping*.) This has less relevance today, but in ancient times walking in pace would have helped our ancestors become aware of the pursuit of a nasty stalker or a saber-toothed tiger.

Behaviors that bring a positive survival advantage will become more common within a species. This can also happen if animals (including humans) experience the beneficial behavior as stimulating or enjoyable, which produces a flow of the "reward molecule" dopamine in the brain. Individuals – and families – that responded to rhythmic behavior by producing dopamine would have, thus, been more likely to walk in step. On the other hand, less rhythmically gifted persons – i.e., those that were bad at paced walking – could have literally stomped themselves out of the genetic pool by discovering the stalker or tiger too late!

Rhythmic behavior might have stimulated the production of dopamine in safer surroundings as well. Clapping hands, stamping, howling around the campfire ... from these activities, the step to dance and music was probably small. Similar mechanisms may have increased the ability of early mankind to perceive, recollect, and mimic sounds. Dopamine certainly flows when modern people listen to music.

FALK: Can you speculate about why the typical gaits of humans (bipedal walking) and chimpanzees (quadrupedal knuckle-walking) affect the development of fetal/infant rhythmic abilities so differently? In other words, why humans universally dance, keep time to a musical beat, and address their infants in musical speech (motherese), but chimpanzees and the other great apes do none of these things.

LARSSON: The brain is formed by early sensory impressions. Newborns, for example, remember the smell of their amniotic fluid. A pregnant mother's diet affects her gestating infant's taste preferences. The mother's voice is also heard in the womb and makes similar impressions. Many people believe that the sounds of mothers' heartbeats were the origin of music. A mother's heartbeat is heard about 27 million times during a pregnancy, so it is an appealing hypothesis, although it has flaws. Our closest relative, the chimpanzee, is totally unmusical despite a heart that sounds just like ours. Something that separates us from the great apes, however, is our two-legged walking style. Chimpanzees can walk on two legs but rarely do so. They move in other ways, and their knuckle-walking and tree climbing create weaker, uneven rhythms compared to the bipedal gait of humans.

Heart sounds stimulate a single sense – hearing. But when the mother walks, it stimulates not only the fetus' hearing, but also its balance, feeling, and proprioception (the sense that reads the positions of the joints) (figure 9.3). Bend and stretch, bend and stretch, swing, swing, everything the baby senses is synchronized with the mother's steps! I think this is good training for keeping time to a musical beat. In other words, we all got a solid basic course in dance when we were in the uterus.

The rhythm of the normal walking pace for humans is around 120 beats per minute. The same pace is common in pop, jazz, folk, classical, and many other kinds of music. The heart beats about 70 times per minute – slower than almost all music. The chimp fetus is also moved when its mother knuckle walks, but in a much more irregular tempo that is usually far from the normal pace of human music. Bipedal gait is not exclusive to humans, but humans are the only *habitual* bipedal primate. Chimpanzees and bonobos, genetically the closest species to humans, essentially lack the ability of complex vocal learning and are unable or almost unable to tap in synchrony with other individuals. I think their irregular locomotion, and the associated locomotion sounds

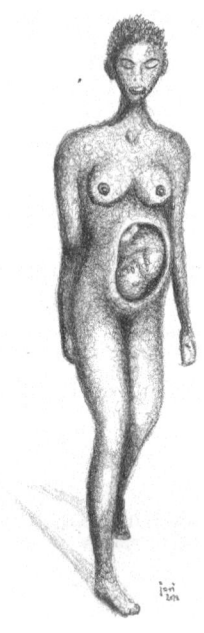

Figure 9.3. A pregnant woman. The neurological development of fetuses that contributes to their eventual acquisition of language occurs partly in response to the rhythmic sounds they hear in the womb, including the cadence of their mothers' footsteps. Reproduced from Larsson et al. 2019 with permission from Joachim Richter.

that chimpanzee fetuses hear, may be key. The ability to keep time to a musical beat is also likely to influence the human newborn's ability to solve certain problems embedded in language.

FALK: **Many scholars think that the musical, rhythmic qualities of baby talk (motherese) help infants learn their languages. Do you think that exposure to the sounds and movements of bipedal maternal footsteps by hominin fetuses (and postnatal infants that were carried by their mothers) may have contributed to the emergence of motherese in addition to the other musical behaviors you discuss? If so, is it possible that the unique prosodic**

aspects of language itself might, ultimately, trace back to the footfalls of prehistoric mothers as hominins made the transition to bipedalism?

LARSSON: Studies suggest that the child-directed speech (motherese) of humans helps children identify where words begin and end. It plays an important role in language learning, engaging infants' emotions and highlighting the structure of the language they hear, and it may help babies decode the puzzle of syllables and sentences. In essence, motherese is like an exaggerated and somewhat musical form of speech. Therefore, I believe that early musical training, such as "the human fetal dance course," as well as bipedal carrying of newborns, is a perfect introduction to motherese and, thus, to humans' lifelong training in language.

To summarize, I think the transition to bipedalism was crucial for the evolution of human music and rhythm, and that these abilities were mandatory for complex vocal learning and the evolution of language. Notably, we have two important but very different mechanisms to explore further if we want to know more about how we became the rhythmic primate. First, the acoustical advantage hypothesis proposes a mechanism in the *phylogenetic* development of musical abilities (i.e., hominids that tended to walk in pace could detect important signals in the surroundings, such as sounds from stalkers, and survive). That process had the potential to transform the human genome. On the other hand, bipedal stimuli in utero and from being carried by bipedal parents may have contributed to the *ontogenetic* development (i.e., the individual's) of rhythmical abilities from the time of fertilization of the egg through adulthood.

Which of these two "bipedal mechanisms" is most important? I don't know, but I think both are important and that they mutually reinforced the evolution of human music and language.

2. Kathelijne Koops

The Night Shift

Dr. Kathelijne Koops (figure 9.4) is a professor in the Department of Evolutionary Anthropology at the University of Zurich in Switzerland. Her research focuses on the evolution of tool use and culture. She has studied chimpanzees in West and East Africa, as well as the chimpanzee sister species, the bonobo, in Central Africa. Below she discusses her research on chimpanzee ground nests in the Nimba Mountains of Guinea, which has surprising implications for understanding when, how, and why early hominins shifted from building sleeping nests in trees to constructing them habitually on the ground.

Koops' research on ground-sleeping nests is immensely important for thinking about hominins' transition from sleeping in trees to sleeping habitually on the ground. Remarkably, her research suggests that ground sleep might have already been practiced in some populations of pre-Homo erectus hominins.

> FALK: **How common were ground nests compared to tree nests in the chimpanzees you studied in West Africa, and what methods allowed you to discover that most of the ground nests had been occupied by males? What materials do chimpanzees use to construct ground nests?**

Figure 9.4. Dr. Kathelijne Koops in the field. Photo courtesy of Kathelijne Koops.

KOOPS: At my study site in the Seringbara region of the Nimba Mountains, the chimpanzees make up to 20 percent of their nests on the ground. This is very unusual, since ground nesting is rare in most chimpanzee populations. Chimpanzees generally sleep in arboreal nests. Every weaned ape builds a new nest, or bed, to sleep in every night, and sometimes during the day as a place to rest. The Seringbara chimpanzees make elaborate night nests on the ground, as well as simple day nests. Since the chimpanzees in Nimba are not yet used to the presence of humans, we were not able to observe them, so we analyzed DNA from shed hairs in the ground nests to investigate the identity and sex of the nest builders. We found that ground nesting is a heavily male-biased behavior, especially with regard to elaborate night nests on the ground. Chimpanzees use a variety of plants to build ground nests – as many as 55 plant species have been recorded. They bend and break off terrestrial herbs, as well as woody plants and small trees, to make a comfortable sleeping platform.

FALK: **You have observed that ground nesting can occur in chimpanzees despite the presence of terrestrial predators and the fact that apes do not use fire. What implications does this have for rethinking the received wisdom that the shift to terrestrial sleeping occurred in *Homo erectus*? Might it have begun in earlier hominins?**

KOOPS: In the Nimba Mountains, leopards are indeed present. Hence, our findings on chimpanzee ground nesting suggest that the presence of predators in paleohabitats of early hominins did not preclude sleeping on the ground. The fact that we see wild chimpanzees making nests on the ground shows us that a fully terrestrial lifestyle, first seen in *Homo erectus*, is not in fact necessary for habitual ground nesting to occur. This in turn suggests that ground sleep may have already been practiced in some populations of pre-*Homo erectus* hominins. For *Homo erectus*, sleeping on the ground may have been easier with the use of fire to protect against large terrestrial predators. But overall, the findings on chimpanzees suggest a more gradual transition from tree to ground sleep in the hominin lineage.

FALK: **Do you think that ground sleeping may have been important for hominin cognitive evolution? Why or why not?**

KOOPS: All great apes build nests, which suggests that their common ancestor was also a nest builder. A lot of research has focused on figuring out the exact function of (arboreal) nest building. One hypothesis is that sleeping in a nest is more comfortable than sleeping on a branch, which might have influenced the type and quality of sleep of great apes. Sleeping in a nest on the ground may have a number of additional benefits compared to building a nest up in the canopy. Sleeping on the ground is more stable than sleeping up in the trees, since it removes the risk of potential lethal falls. A more secure sleeping platform, in turn, would have allowed for more intense deep REM-dominated sleep. More comfortable rest and higher-quality sleep would have conferred several cognitive benefits for ground-sleeping hominins, such as enhanced long-term memory consolidation. Hence, ground sleeping may well have played a role in hominin cognitive evolution.

3. Susanne Shultz

Nocturnal Predators – The Downside of Sleeping on the Ground

Dr. Susanne Shultz (figure 9.5) is a professor of evolutionary ecology and conservation biology at the University of Manchester in the United Kingdom. Her graduate work focused on the relationship between primate behavior and African crowned eagle predation, and she has investigated the poisoning of Asian vultures, brain evolution in mammals,

Figure 9.5. Dr. Susanne Shultz at a wolf research center. Photograph courtesy of Susanne Shultz.

and hominin evolution in response to climate change. Her recent work has focused on animal social networks and health, the impacts of anthropogenic environmental change on species range collapse, and disease transmission at the human–wildlife interface. Her core interests can be summed up as understanding the origin of human sociality and cognition and mitigating our negative impacts on the natural world.

In the following interview, Shultz points to evidence in the fossil record that raptors and leopards preyed upon australopithecines. She also discusses the likely impacts that these and other predators had on the sleeping habits, alarm calls, weapon development, and social evolution of early hominins.

FALK: **Based on your studies of predator/prey relationships in certain African species, do you think that predation may have been a serious threat to early African hominins who began spending nights, as well as days, on the ground? What animals are likely to have preyed upon our early predecessors? Would they have been nocturnal, diurnal, or both?**

SHULTZ: Early hominins had no natural defenses such as large canines, robust bodies, or even fast running speeds, so they would have been vulnerable to a range of predators. From studying living primates, we know that hominins that spent most of their time on the ground experienced much higher rates of predation from terrestrial predators such as cats and large snakes, and also from large eagles. Although large cats like leopards and lions were obvious predators, early hominins – especially youngsters – were likely vulnerable to raptors. These risks would have been even greater in open country, where there were fewer opportunities for escape. Early hominins were undoubtedly diurnal, although their predators would have been both diurnal and nocturnal. In open country, this would mean either returning to a safe refuge or accepting high risk. Baboons, which can inhabit the open woodland environments that early hominins moved into, return to safe refuges (such as cliffs or large trees) for sleeping. Their daily paths are constrained by their need to return to these safe sleeping sites each night. This leads to changes in foraging and social behavior. For example, it is likely that hominins aggregated in larger groups around safe places at night and then may have broken into smaller foraging bands during the day.

FALK: **Is there any direct evidence in the australopithecine fossil record that speaks to predation on early hominins?**

SHULTZ: The cause of death for individuals found in hominin assemblages remains highly speculative. Sites with remains from multiple individuals have attracted much attention and demand an explanation. Leopards have long been thought to be the key predator on early hominins. Accumulations of bones in underground caves have been ascribed to leopards stashing prey in trees and the remains eventually dropping into sinkholes. There is also clear evidence for leopard involvement in some hominin deaths. For example, the juvenile australopithecine SK 54 from the site of Swartkrans in South Africa has puncture wounds that match the canines in the lower jaw of a fossil leopard from the same deposits. Perhaps more surprisingly, there is also evidence of predation on young hominins by large eagles. My earlier work highlighted patterns of damage made by crowned eagles on monkey skulls that were very similar to damage found on some early hominin skulls. Specifically, damage caused by eagle predation is extremely similar to the damage on the skull of the Taung child, a young australopithecine. However, in most cases the cause of death is unknown, partly because the traces left by predators such as bite marks or scratches are often lost in the process of preservation. The bones left behind are frequently broken fragments, which may have been caused by a predator or scavenger or could simply be the result of deposition and decay. It is likely that we can make better inferences about the causes of hominin mortality if we carefully study how different predators use their landscape and damage prey remains.

FALK: The hominins that began constructing terrestrial sleeping nests are thought to have been around the same size as living chimpanzees. They would not yet have occupied home bases, made fire, or invented language. If the threats from predators were as severe as some think, what other kinds of behavioral changes might have helped them survive?

SHULTZ: Living primates give us some clues about the past. Species that live on the ground are found in larger groups than those that live in the trees. Those that live in open habitats are found in larger groups than those that live in closed forests. These species often use mobbing and counterattacks to deter predators. Baboon and chimpanzee males jointly counterattack potential predators. Both species return to safe sleeping sites: chimpanzees construct nests in trees, and baboons return to sleeping trees or cliffs. So, we can predict that early hominins likely lived in fairly large social groups that used some sort of refuge for sleeping. Like chimpanzees and baboons, they likely used some sort of communal defense or mobbing. Although they would not have had fully developed language, they undoubtedly had vocalizations that warned group members about risks. We can also infer this from living primates, who use alarm calls that contain a lot of information. For example, some African and Asian monkeys use distinct calls for ground versus aerial predators. Variation in the intensity and duration of the calls conveys information about relative risk. They also modulate the use of calls for ambush versus pursuit predators. Alarm calls are most effective for discouraging ambush predators that rely on surprise but are less effective for pursuit predators. Some behaviors, such as the use of weapons, are harder to infer. Chimpanzees use sticks and rocks in displays but have also been seen to use them to attack models of predators. Human pastoralists often carry spears to protect themselves and their livestock from predators. Given the gracile bodies of early hominins, their relatively slow running speeds, and their lack of natural weapons, it seems highly plausible that early hominins used simple weapons such as sticks and spears to defend themselves – but there is no direct evidence for this. We also do not know when hominins started to modify their environments to make safe spaces, such as using thorny branches to construct bomas or sleeping compounds. This probably did not happen until much later in hominin evolution when we start to see the use of fires and hearths, but earlier use of protection would not have been preserved in the fossil or archaeological record.

4. Cara Wall-Scheffler

Women – The Burden-Bearing Sex

Dr. Cara Wall-Scheffler (figure 9.6) is a professor of biology at Seattle Pacific University. Her research focuses on the physiological ramifications of different body shapes and sizes in living and extinct hominins. Some of her experiments have implications for how prehistoric women may have moved and carried babies.

As you will see from her interview, Wall-Scheffler's scientific experiments provide evidence for her conclusions about why women are better adapted for carrying heavy burdens than men. She also reasons that baby slings were probably the first "invention" associated with becoming a habitual terrestrial biped.

Figure 9.6. Dr. Cara Wall-Scheffler with a student. Photograph courtesy of Cara Wall-Scheffler.

FALK: **You have observed that women around the world who live in hunting and gathering communities are "the burden-bearing sex." What kinds of loads do women carry and is there an evolutionary explanation for why women, rather than men, do the heavy carrying? What kinds of experiments have you done that shed light on this matter?**

WALL-SCHEFFLER: Women carry every kind of load imaginable. Besides loads related to being pregnant and carrying babies after they are born, women carry water, wood, and food, in addition to other kinds of household items (baskets, barrels, tools, and pots) and personal items. Since women's load-carrying is a human universal, it is especially interesting from an evolutionary perspective. Based on evidence about the physical forms of bodies, hominins of both sexes probably participated in carrying loads prior to *Homo sapiens*, but there does seem to have been a shift with us that conferred the ability to carry loads on females in particular. Since there is one load that is non-negotiable for population survivorship (female pregnancy), it seems reasonable that females would maintain excellent load-carrying capabilities throughout the life span, regardless of other changes in lifestyle that we see in *Homo sapiens*.

My collaborators and I have approached this topic primarily by measuring variables that relate to effective movement (locomotion) – usable energy, muscle contractions, speed, fatigue – as well as measuring the related variation in the physical form of the body. We found that not only are women excellent walkers who use less energy than men during all walking tasks (on average), but they also have the same efficiency as men when carrying loads that are relatively more challenging (a larger proportion of their body size). We have also shown that the measurements that people often claim are different between females and males (such as torso breadth) are actually widely variable around the world and have nothing to do with giving birth! But, of course, these measures are hugely important for carrying loads.

FALK: **What are your thoughts about the origin(s) of baby slings?**

WALL-SCHEFFLER: I have always argued that baby slings would have been the first "invention" of becoming a habitual terrestrial biped. Though some people today balance their child on their shoulders, it would have been much more effective for walkers to sling infants close to the waist. I have shown that carrying children on the back, in particular, is one of the most effective and efficient ways to carry children. I think a key question now is: How simple can a sling be and still save substantial energy for the carrier? We know quite a bit about slings around the world, and when I say simple, I mean just a single piece of material. For example, many people who live in the Amazon use a single, thin strap of material – kind of like a jump rope. The strap goes diagonally over their body, and a child sits in the loop that is created at the opposite hip. The child either has to brace itself against the carrier's body, or the carrier has to brace the child to keep it from toppling right out of the loop. In this case, the weight of the child is not in the arms of the carrier, but the carrier's arms are still needed to stabilize the child. It is an open question how much energy this might save compared to slings that are substantially wider or have multiple straps that can support the child's bottom and back, freeing the carrier's arms and hands from any holding or supporting functions.

FALK: **Has your research influenced how you personally exercise or carry things?**

WALL-SCHEFFLER: I know that walking while carrying something uses a lot of energy, so I'm quite happy to walk to the grocery store instead of going for a jog once a week. Carrying things also uses different muscles, so I like to diversify my workouts in this way, and I vary whether I take a backpack or carry my groceries home in my hands. Walking slowly – specifically when I am running errands (going to a specific place that is a specific distance from my home) – uses the same amount of energy as walking to that place quickly; it's just the time it takes that is different, not the energy. This means that when

I'm walking with my dog or with my child and they are dawdling, even if I don't feel like I'm using a lot of energy, I actually am! I use this knowledge to try to be more patient with my family's pace.

5. Helen Anderson

Crosshatching, Trellises, and Diamond Patterns

Dr. Helen Anderson (figure 9.7) is a curator in the Africa section of the Department of Africa, Oceania and the Americas at The British Museum in London. She previously worked as a Research Officer at the Rock Art Research Institute at the University of the Witwatersrand in Johannesburg, South Africa. She has researched and published papers on the origins of art, art and neuroscience, African rock art, historical photographs, and basketry. Below, she discusses her research on the origins of basketry – research that has surprising implications for understanding the evolution of advanced cognition in humans.

After this interview, Anderson's theory that geometric designs "reflect a familiarity with patterns existing in the visual repertoire of the maker" inspired me to investigate the mystery of why geometric figures are always earlier than representational ones in the archaeological record of art – known as the "geometric enigma" (Falk 2024).

> FALK: **What is the earliest sign of mark-making you know of in the archaeological record, and how early did it occur? Do you think there could have been an evolutionary connection between the weaving of arboreal sleeping nests in the common ancestor of chimpanzees and humans and the emergence of basketmaking in prehistoric hominins?**

Figure 9.7. Dr. Helen Anderson preparing to repair a basket. Photo courtesy of Helen Anderson.

ANDERSON: I think mark-making is probably an activity that goes back hundreds of thousands of years in our cognitive evolution. One of the earliest examples that interests me is an engraved freshwater shell from Java, Indonesia, thought to be made by *Homo erectus* and dating back to 500,000 years ago. Incised with a zig-zag pattern, possibly using a shark's tooth as a tool, the engraving would have produced white lines on the brown outer shell, resulting in a striking pattern. Neuroscience has repeatedly shown that pattern recognition is the fundamental basis for understanding cognition in humans and that related capabilities became increasingly sophisticated during the evolution of the cerebral cortex.

Nest building is a feature of all three great apes, and it seems likely that this behavior was a feature of their last common ancestor who lived around 14 million years ago. There is growing evidence that great apes have some understanding of the materials they use and how they use them – for example, orangutans choose branches with greater rigidity for the main structure of their nest with thinner branches for the lining, and chimpanzees modify their nests to adapt to current weather conditions. Contemporary human use of plant-based materials for creating shelters and bedding is a practice that developed during hominin evolution, and the capacity to weave may indeed have developed from such practices. Also, humans and our predecessors could have learned to weave though emulation. Neuroscientific research in the last 20 years or so on mirror neurons has shown just how well humans learn by imitation.

Basketry, in all its various forms, is characteristic of our species. There are few cultures who have not used basket-weaving skills for fundamental purposes related to shelter, protection, harvesting food, storage, and so on. I think in evolutionary terms, the capacity to weave developed our cognitive abilities related to numbers, patterns, and structures. Basketry may, indeed, have had a longer and more meaningful place in human culture than previously considered and may have played an important role in the evolution of wider patterns of cognition and understanding.

FALK: You have written that archaeological evidence from Eurasia shows that advanced basketry and loom-woven textiles date to around 30,000 years ago and that rope and twine were produced as early as 40,000 years ago. What

kinds of evidence do you use to infer the even earlier dates of baskets and other textiles such as nets?

ANDERSON: By the time fiber-based technologies such as cordage, basketry, textiles, and netting appear in the archaeological record during the Upper Paleolithic in Europe, they do so fully formed, which suggests that their origins may be considerably older. But how might we find evidence for the antecedents of these technologies? One avenue is to consider the tendency to transfer weaving designs onto other materials and objects, which we see in the Upper Paleolithic in the form of geometric patterns such as crosshatching, trellis, and diamond patterns on artifacts like adzes, needle cases, personal ornamentation, figurines, and ostrich eggshells. I think we can detect the same inscribing practices even earlier in the archaeological record and still make connections with the processes of weaving and basketry. Crosshatched designs on stone, ocher, and ostrich eggshells have been found at a number of sites across southern Africa dating to as early as 100,000 years ago. In the absence of direct archaeological evidence for basketry or netting, these designs suggest that fiber-based technologies may have their antecedents here.

FALK: Do you think these kinds of inscribed geometric patterns were evidence of purely symbolic and abstract thought, as some have suggested? If not, is there an alternative explanation that focuses more on the practical experiences of the makers?

ANDERSON: The crosshatch, trellis, or diamond-like patterning on stone, ocher, and ostrich eggshells represent what is often considered evidence of symbolic thought or even the emergence of "art." There is similarity and recurrence in the designs found at sites that are hundreds of kilometers and thousands of years apart. The significance of these incised artifacts for many archaeologists is that they are abstract designs, unrelated to reality, and as such used as proxies to indicate symbolic thought and potentially the emergence of language to convey meaning. However, very little attention has been placed on the actual patterns themselves or why they are thought to represent the first evidence of what we might term "symbolic behavior." I am more inclined to see this motif as being meaningful because it has roots in a shared visual and cultural repertoire that is based on lived experiences.

Mounting archaeological evidence indicates that after 100,000 years ago, modern humans were engaging in sophisticated fishing technologies likely to require the construction of some forms of nets, traps or weirs, or containers made from plant material to both catch and transport the fish back to cave sites. Hunting probably involved nets or traps as well. Crosshatch patterns have been found on hundreds of fragments of ostrich eggshells, thought to be the remnants of ostrich-egg water containers, which were likely carried in some form of netting – a practice continued to this day by the Kalahari San and evidenced in rock art images in southern Africa. For me, these designs reflect a familiarity with patterns existing in the visual repertoire of the maker. The design was not an abstract concept; it referenced cordage, thread, nets, traps, and woven containers. It was meaningful through a lived emotional experience, the acquisition and transportation of food and water.

6. Rebecca Biermann Gürbüz

Woodworking and Cognition

Dr. Rebecca Biermann Gürbüz (figure 9.8) is an anthropologist based in northwest Florida. She holds a PhD in anthropology from the University of Buffalo and worked as a professional archaeologist for a number of years. In her academic research, she uses experimental archaeology and computational techniques to study the evolution of hominin behaviors, and has recently coauthored a thought-provoking paper with Dr. Stephen Lycett titled "Could woodworking have driven lithic tool selection?"

Gürbüz's experimental research suggests that the earliest stone tools in the archaeological record were likely used not just for butchering animal carcasses, as traditional views suggest, but also for making botanical tools — a.k.a. woodworking.

FALK: **Which do you think came first, woodworking or stone tools? What kinds of primatological or ethnographic evidence shed light on the origins and prehistoric functions of woodworking?**

GÜRBÜZ: Based on the fact that all great apes (as well as some other nonhuman primates like capuchin monkeys) modify plants to find food such as ants and termites, it's logical to think that our shared ancestors also modified plants to acquire food. In this specific way, woodworking almost certainly preceded the origin of stone tools. However, if we define woodworking as using another tool (not teeth or hands) to modify wood, then obviously that requires the invention of tools, such as stone tools, first.

Figure 9.8. Dr. Rebecca Biermann Gürbüz exploring an archaeological site in Turkey. Photo courtesy of Rebecca Biermann Gürbüz.

Chimpanzees use plant-based tools in nut-cracking, spear-assisted hunting, ant-dipping, termite fishing, and digging up underground foods. In order to create spear-like tools to hunt bush babies, chimpanzees in Fongoli, Senegal, also sharpen wooden sticks. Other, more distantly related non-human primates also make tools from plant materials. For example, wild orangutans remove leaves and bark from branches to harvest insects and seeds, though this behavior seems less common than in chimpanzees. Even wild mountain gorillas use tools, though it seems even more rare than in orangutans.

Taken together, the evidence from all of the great apes (chimpanzees, bonobos, orangutans, and gorillas) indicates that modifying plants to create simple tools is a behavior that was likely practiced by their common ancestor as well as early human ancestors.

FALK: **What experimental research have you done that explores the impact woodworking may have had on the invention of early stone tools? What did you find?**

GÜRBÜZ: While many archaeologists argue that butchery was the key behavior that led hominins to invent and develop stone tool technology, it's also possible that woodworking may have played a role, particularly when you consider the evidence for hominin plant modification. This is where our experimentation comes in.

In addition to the primatological evidence discussed above, which suggests that our hominin ancestors were almost certainly modifying plants to create simple tools, there is also archaeological evidence that shows that once stone tools were invented, they were used in woodworking. The clearest evidence for woodworking comes in the form of wooden spears. While it's very likely that humans were making wooden spears before they appear in the archaeological record due to the perishable nature of plant materials, the oldest spears are about 400,000 and 300,000 years old, from England and Germany. Archaeologists have also found secondary evidence of woodworking on stone tools, namely, plant residues and woodworking-related wear on stone tools. The oldest of these tools date to 2 and 1.5 million years ago and were found in Kenya and Tanzania.

Despite the primatological and archaeological evidence for hominin woodworking, most scholars only look to butchery as the key behavior that led to the invention of stone tools.

A number of archaeological experiments have demonstrated that tool shape and size are linked to how quickly and well a task can be completed. Hominins would likely have wanted to complete tasks as efficiently as possible, and so Dr. Stephen Lycett and I designed an experiment specifically to test how well very early stone tools can complete a woodworking task. Flakes are the oldest known stone tools in the archaeological record, as well as the simplest. They are akin to sharp, simple knife blades. The reason we looked specifically at stone tool *size* is that several experiments have shown that bigger tools are generally better at cutting tasks, both general cutting and butchery. Furthermore, the earliest stone tools in the archaeological record, from Lomekwi, Kenya, are notably large. Our experiment shows that large flakes are significantly more efficient during woodworking compared to small flakes. Most scholars associate stone flakes exclusively with butchery, even when an archaeological site doesn't have any other evidence of butchery (like cut-marked bones). Our results indicate that these early flakes could also have been created and used for woodworking.

To be clear, there is archaeological evidence for butchery around the same time as the earliest stone tools (3.4 and 3.3 million years ago, respectively). Our results by no means indicate that stone tools were not used for butchery; however, they do indicate that early flake tools, as well as later technologies, could have been created and used for woodworking in addition to butchery.

FALK: Do you think the invention of bifacial hand axes was associated with a "cognitive leap" as some archaeologists suggest? Is it possible that the invention(s) of wood tools

was equally important for hominin evolution? If so, which one(s) and why?

GÜRBÜZ: What is clear is that the production of bifacial Acheulean hand axes requires a specific set of cultural and knapping behaviors that was not required for stone flakes, the simple tools that preceded them. Whether or not this constitutes a leap or a gradual progression that may or may not be archaeologically detectable is unclear.

Around the time that the Acheulean ended (about 500,000 to 300,000 years ago), the earliest direct archaeological evidence for wooden tools started to emerge – wooden spears. However, it's likely that hominins were creating and using a variety of plant-based tools, including rough, spear-like, sharpened sticks, much earlier (at the time of their split with chimpanzees). Refined, sharp spears created with stone tools are a much younger invention than the older, rougher version. While there is earlier secondary evidence for stone projectiles that could have been attached to wooden spears, the earliest direct surviving evidence for wooden spears dates to 400,000 and 300,000 years ago in England and Germany. At the site in Schöningen, Germany, 10 spears have been found, making it the biggest early spear site. The most remarkable thing about these spears is that their tips are slightly off-center, avoiding the softer center of the wood and ensuring that the tip was created from the hardest wood. Several experiments have shown that the Schöningen spears are weighted like modern javelins and would have been effective throwing spears.

The evolution of throwing spears demonstrates the advancement of hominin technology, which reduced the dangers of hunting by increasing the distance between the hunter and the prey. While the earliest spears very likely predate the specimens identified in the archaeological record, these early artifacts nonetheless provide evidence for advanced cognitive reasoning.

7. Glenn Marshall

Rafting to Australia and Flores – How'd They Do It?

Glenn Marshall (figure 9.9) is manager of sustainable infrastructure projects in Alice Springs, Australia. He has degrees in geology and chemistry, environmental science, and a partly completed graduate diploma in archaeology. You can read about his adventures at his website: https://www.chasing-archaeology.com/home.

I met Marshall in 2019 at the Asia Pacific Conference on Human Evolution in Brisbane, Australia, where he presented a fascinating poster about an upcoming experiment he and his colleagues were preparing. We got to talking about a mystery in Southeast Asian archaeology – namely, the question of how the predecessors of tiny LB1 and her kind (Homo floresiensis) got to the island of Flores over a million years ago. After the conference, my family and I visited Marshall in Alice Springs, where I asked him if he thought LB1's predecessors might have been lodged in sleeping nests in trees that were uprooted and blown out to sea during a tsunami or cyclone. When he didn't rule the idea out, I decided to research and write a book that would explore the roles played by sleeping nests, baby slings, and other botanical inventions during the cognitive evolution and global migration of our earliest ancestors. Five years later, The Botanic Age *is the result of that conversation!*

Figure 9.9. Glenn Marshall with current drifters that he built.

FALK: **Who are the First Mariners and what was the goal of their recent project?**

MARSHALL: The First Mariners is a group of experimental archaeologists that includes ocean sailors, stone tool experts, maritime historians, and early human migration enthusiasts, all united in our common love of adventure, curiosity, and practical experimentation. We build and test watercraft like the ones humans might have used to cross sea straits to colonize new lands. The goal of our recent project was to explore how people first arrived in Australia around 65,000 years ago. We knew they must have island-hopped through

Figure 9.10. Completed raft for Marshall's 2020 project.

the 1,000 km-wide Indonesian island chain, and that they became the world's first great maritime culture (that we know of). We decided to build and test a large bamboo raft with a small sail (figure 9.10), voyaging from the most southern island of Indonesia (Timor) to the now-submerged northwest shelf of Australia. We launched during the summer monsoon season when winds blow from the northwest for a few months (the ideal crossing time). We wanted to find out if a bamboo raft tied with local palm fiber, carrying 10 crew, being pushed by wind and current, could cope with the crossing and end up on the Australian coast. The villagers of Oeseli, on the nearby island of Rote, brought to our project thousands of years of knowledge about their seas and their experience building vessels from bamboo and fiber rope. My main role was to build and launch multiple ocean current drifters to better understand the direction, speed, and influence of the strong currents and winds that flow through the islands, and to monitor their impacts on the raft as it voyaged.

FALK: What happened and what did you learn?

MARSHALL: After finishing the raft and conducting successful sea trials, the Indonesian Department of Immigration declined to issue exit visas to sail into international waters, citing safety and visa concerns. We adopted Plan B, towing the raft offshore and releasing it without a crew. This started a remarkable 1,600 km unmanned voyage, dictated by winds and currents, in keeping with what we had learned about local currents and winds. A tracking device I made for the raft revealed that it traversed the Timor Sea, survived a cyclone, and came within 60 km of the Australian mainland. If we were onboard, we would have made landfall. The winds then reverted to dry season conditions and pushed the raft 800 km back to Timor. A remarkable voyage, indeed! It showed that the mere 80 km water gap between Timor and Australia that existed around 65,000 years ago (much smaller than today's water barrier) could have been safely traversed within a few days during suitable weather. People who crossed could return when the seasons changed, facilitating early short-term exploration voyages and return voyages once new islands were settled. A raft of our size was probably not the vessel used by Australia's first colonizers – it was too large, too heavy, too specialized – and there is not yet evidence that people used sails at that time. But our venture provided important real-life data and food for thought for scientists who model such crossings, and for archaeologists who excavate remains on those islands. I'm inclined now to think people used their near-shore, day-to-day watercraft (most likely small canoes) to cross new straits during favorable weather, with repeat crossings over several years to move enough people across to form sustainable populations.

FALK: **What's next?**

MARSHALL: Importantly, the project generated great pride and interest from people in Oeseli. I am planning a new project with them, to paddle their small canoes across significant straits, starting at 10 km and working up to the same 80 km crossing that our raft achieved within the first two days of its journey. For years, I've been interested in how hobbits arrived on Flores, and available evidence points to accidental rafting from the north, not deliberate seafaring. But what sort of rafts? (I'm guessing tangled trees from a landslide.) What islands were they swept off? (Probably Sulawesi.) Were local landforms, sea levels, currents, and monsoonal conditions the same a million years ago? (It seems that roughly they were.) How many days to float to Flores? (Up to seven.) We aspire to construct a few "accidental rafts" from local trees, put proxy-hobbits onboard, launch them into the sea, and track them toward Flores.

8. Louie García

The Fiber Arts, Then and Now

Louie García (Tiwa/Piro Pueblo) (figure 9.11) is a traditional fiber artist whose work has been exhibited in many museums. He is a member of the Cedar Mesa Perishables Project, a team of archaeologists and Pueblo

Figure 9.11. Pueblo fiber artist Louie García.

weavers that is documenting prehistoric perishable collections in institutions across the United States. Louie has a BA in biology and Spanish, as well as an MA in language literacy and sociocultural studies.

García's interview beautifully illustrates that whispers of the Botanic Age remain today, not only in the continued importance of wood, paper, and textiles for practical reasons, but also in the use of botanical materials and weaving for creating artistic pieces and reinforcing spiritual values.

FALK: **Who taught you to weave? What kinds of skills and materials are used to create fiber art? Are aspects of weaving ever meditative for you?**

GARCÍA: I learned the basics of Pueblo weaving from my maternal grandfather. Since then, I have conducted much of my own research and have studied both public and private collections to learn more about both prehistoric and historic Pueblo and Basketmaker textiles and techniques. Pueblo fiber art techniques and materials are quite varied. There are many techniques and materials that have fallen out of use with time but there are also some techniques and natural materials that have changed little since ancient times. Materials used to produce textiles over time in the Pueblo southwest include yucca fiber, dog hair, human hair, Indian hemp, turkey feathers, rabbit fur, and cotton. Sheep wool was introduced by the Spanish during the colonial period and is still used today in many Pueblo textiles.

Each step in the process of Pueblo weaving is meditative if meditation is looked at as a form of prayer. Everyday activities such as getting up in the morning begin with prayer as we greet the sun and thank him for another day. Much like this, everyday tasks are usually preceded by prayer, whether you're sitting down to a meal or planting a garden. In this way, we acknowledge our ancestors and the spirits that we, as Pueblo people, believe in. As I sit at my loom, I am often singing a song in prayer as I weave good thoughts and good intentions into each weaving that will be used by an individual or for use in a ceremony to bring good things to the people.

FALK: **You have said that you remember your ancestors through weaving. How ancient do you think weaving is and in what ways does it help you remember your ancestors?**

GARCÍA: I always understood that Pueblo weaving is one of the most ancient of Pueblo art forms. I always had a fascination with the beautiful textiles used in the dances and various ceremonies that go on in our Pueblo communities. Through my research and collaboration with various archaeologists, I have learned a great deal about the antiquity of our Pueblo weaving tradition as well as my own Pueblo ancestors. Some of the earliest textiles found in the Pueblo southwest date back two thousand years. The fact that the Pueblo fiber arts tradition is that old, and that some of the tools and techniques have remained unchanged for over a thousand years, gives me a strong sense of connection with my ancestors. The Pueblo people are farmers, and along with growing food, cotton has also been an important cultivar in many Pueblo communities for several centuries for religious use. There is an unbroken chain of Pueblo traditions that are still maintained in many of our communities today and we honor that connection by carrying them on and teaching them to our young people.

FALK: **Do fiber arts have symbolic elements that have been passed down through generations through oral histories or in origin stories? If so, can you share an example?**

GARCÍA: There is much symbolism behind each of our Pueblo textiles. All of our textiles are left as they are woven on the loom. They are not cut or tailored in any way as they are woven to the size needed. We are most proud of our handwoven textiles, since many of the Pueblos embroider on commercial cotton monk's cloth, and handwoven cloth today can be difficult to find and pricey. Since most of our textiles are made of cotton, as they were over a thousand years ago, and since we live in a desert with little rainfall, clouds and rain are two very important elements in Pueblo culture. As agricultural people, we depend on the rain and moisture for our crops to mature so that we can provide for our families. For us, cotton represents clouds. Therefore, our cotton textiles are essentially clouds that we dance with as we pray for moisture in our ceremonies.

A Final Word

These interviews provide brief glimpses into some of the research and creativity that inform the ideas presented in *The Botanic Age*. You can find out much more about what's behind the book in the following notes for each chapter and in the list of references.

Notes

Preface

1 Margaret Mead (1901–78) did pioneering fieldwork in South Pacific and Southeast Asian small-scale cultures where she studied cultural expectations about adolescence, childhood, and gender. One of her best-known books is *Coming of Age in Samoa* (1928).
2 Henrich, Heine, and Norenzavan 2010, quotation from p. 61.

Introduction

1 Archaeologists divide the Stone Age into the Old Stone Age (Paleolithic) starting with the earliest recognized signs of deliberately made stone tools around 3.5 million years ago; the Middle Stone Age beginning around 300,000 years ago; and New Stone Age (Neolithic) from around 12,000–5,000 years ago.
2 *2001: A Space Odyssey* (1968) was directed by Stanley Kubrick.
3 Lith from the Greek *líthos* means "stone."
4 The feet of bipedal hominin infants (as well as adults) had evolved from grasping organs to weight-bearing ones that supported walking.
5 Tanner and Zihlman 1976; Zihlman 1981, 1985. The research of distinguished emerita professor of anthropology Adrienne Zihlman and Nancy Tanner (1933–89) was truly pathbreaking.
6 This is also sometimes referred to as "Underwater Basket Weaving 101," although the term I (and many of my peers) was familiar with during my university days was "Basket Weaving 101."

Chapter 1. Baskets in the Trees

1. Some question evolution because of incorrect ideas about how it occurs or due to religious beliefs. Such skepticism existed even before 1871 when Charles Darwin first published his ideas about humans in a two-volume set (*The Descent of Man*) – 12 years *after* he published *On the Origin of Species* (1859; frequently shortened to *Origins*). The reason for this delay was because Darwin feared his readers would reject his broader ideas about evolution if he included them in his 1859 magnum opus. By 1871, however, *Origins* had been so successful that Darwin figured he could finally incorporate humans into his grand theory. Although some religious institutions such as the Catholic Church now accept evolutionary theory, many people still harbor misconceptions about how evolution works. College students, for example, may arrive at their first anthropology courses with the preconceived notion that Darwinian evolutionary theory implies people are descended from living apes, which it does not, of course. Contemporary apes are the products of their own long, separate evolutions, and science has shown that people and apes are descended from common ancestors that lived in the very deep past.
2. Some count four kinds of great apes because chimpanzees include both common chimpanzees and bonobos. Bonobos are a distinct species of chimpanzee that have received wide interest partly because they have highly active sex lives. Further, as Groves 2018 opines (p. 21), "it is difficult to escape the conclusion that it is customary to treat the Bonobo separately simply because it has its own common name!" There is also a much smaller ape, the gibbon (*Hylobates*), which is known as the "small-bodied ape" (a.k.a. "lesser ape") rather than as a great ape and is evolutionarily more distant from humans than the great apes.
3. See Groves 2018 for discussion. Dates are estimates from the public knowledge base timetree.org. Some current researchers estimate earlier splitting dates than those shown here for humans and great apes (Kuhlwilm et al. 2016). Timetree estimates the split between chimpanzees and humans to have occurred at around 6.5 MYA.
4. The timing of the evolutionary divergences of the different kinds of primates (or other animals) is estimated scientifically by comparing their genes, and the results may be illustrated on evolutionary trees such as the one shown in figure 1.1. Along with fossils, genetic studies have shown that the hundreds of species of living monkeys are more distantly related to humans than apes, and that they originated at different times during the past 40 million years. Most researchers estimate that the human and chimpanzee lineages diverged at some point (which varies among researchers) between 4 and 8 MYA, based on comparative genetic and/or fossil studies (Grabowski and Jungers 2017; Suntsova and Buzdin 2020). Textbooks often narrow the range to 5–7 MYA. The two oldest putative fossil hominins are *Orrorin tugenensis* (~6 MYA) and *Sahelanthropus tchadensis* (~6–7 MYA) (Brunet et al. 2002). All of these estimates are based on the current state of knowledge and could, of course, change with new discoveries. In this book, an estimate of 6.5 MYA is used for the split between chimpanzees and hominins, per timetree.org.
5. Africa is the only place human ancestors (hominins) lived until some of them began migrating to other parts of the world around 2 MYA.
6. Prieur and Pika 2020.

7 Goodall 1964. Interestingly, Goodall's discovery was preceded by Wolfgang Köhler's (1925, p. 79) observation of ant-fishing in captive chimpanzees: "So, first one of our animals, then another, and then the whole company, began to stick twigs and straws out through the meshes and drew them in immediately, covered with ants, which were promptly devoured." Köhler (p. 112) also describes a chimpanzee (Koko) who, unable to reach an object with a stick, plucked a stem from a geranium bush and plucked leaves from it as he approached the desired object.
8 Pascual-Garrido and Almeida-Warren 2021.
9 Koops 2020.
10 Vančatová and Vančata 2021.
11 Shumaker, Walkup, and Beck 2011, quotation from pp. 109–10. The authors count four great apes because they differentiate bonobos from other chimpanzees.
12 van Casteren et al. 2012.
13 Fruth and Hohmann 1996.
14 Fruth and Hohmann 1996, quotations from pp. 226 and 238.
15 Old World monkeys are from Africa and Asia, whereas New World monkeys come from Mexico and Central and South America.
16 Zihlman and Underwood 2019.
17 Povinelli and Cant 1995.
18 According to Grabowski and Jungers 2017, the common ancestor of the great apes had a body that was approximately the size of today's little gibbons. Presumably, the evolution of more mobile upper limbs allowed this ancestor's descendants to obtain more food, which contributed in turn to the evolution of larger body sizes in the three great apes. Part of the reason "greatness" evolved in orangutans, gorillas, and chimpanzees may, thus, have been related to availability of food such as ripe fruit (Fruth and Hohmann 1996). The common ancestor of chimpanzees and hominins is thought to have been roughly the size of contemporary chimpanzees.
19 Stewart et al. 2011.
20 Fruth and Hohmann 1994, p. 311.
21 Fruth and Hohmann 1994.
22 Anderson et al. 2019.
23 Based on observations of a very young chimpanzee, Köhler (1925, p. 95) ventured that nest-building was one of the very few behaviors that appears to be instinctive in chimpanzees. Grooming was another (see p. 321).
24 Anderson et al. 2019, p. 324. As is often the case in evolutionary biology, looking at nest building through a stark nature/nurture lens fails to capture the fact that it has both biological and cultural components, i.e., it is best viewed as a biocultural phenomenon, somewhat like humanity's linguistic abilities. You weren't born with the ability to read these words; instead, you arrived in the world with an evolved genetic makeup (nature) that enabled you to learn the spoken and written language(s) you were exposed to (nurture, or culture). Just as all populations of humans have language, the daily practice of building sleeping nests is universal across the great apes.
25 Stewart et al. 2011, p. 390.
26 Not all sleeping nests are like this, however. For example, in some cases, nests integrate multiple adjacent trees and are "so flexible that the 'bowl' resembles a hammock" (Fruth et al. 2018, p. 501).

27 van Casteren et al. 2012.
28 McLennan 2018.
29 McLennan (2018, p. 227) observed that "Irrespective of whether nest tying constitutes true knot making – commonly considered absent in wild great apes – this nest construction technique would seem to require advanced dexterity and a sophisticated understanding of the mechanical properties of the plants used." Wolfgang Köhler (1925, p. 324) anticipated McLennan's discovery nearly a century earlier when he described knot tying in a young female chimpanzee named Nueva within days of her having arrived at the Primate Station on Tenerife in the Canary Islands: "She was busy with a woollen rag which she tied to a stick; she was not content with simply wrapping it around the stick, but actually achieved a sort of knot, by looping one end of the rag through the portion wound round the stick and pulling it taut. However humble this effort may seem to the general public, it has an amazingly *constructive* character for anyone who knows the tearing, smashing, and demolishing tendencies of the species. Other apes than Nueva also liked to poke about with straws (or sticks) in holes and crevices, but I never observed any of them 'weaving' and carefully plaiting straws through the wire intersticies as she did. She had a special fancy for knots."
30 Stewart, Piel, and McGrew 2011; Fruth and Hohmann 1994. There is also a good deal of variation in other physical features of sleeping nests, such as how high they are located, what kinds of trees they occupy, etc. (Fruth and Hohmann 1996). These characteristics are associated with environmental factors including rainfall, temperature, nature of the terrain, and the presence or absence of predators, as well as with social variables like the ages and sex ratio of the individuals making up various populations. Females, for example, construct nests that are, on average, higher than those of males (possibly related to safety) and tend to make more day nests (Fruth, Tagg, and Stewart 2018).
31 Interestingly, modern humans also develop an apelike wariness of falling – who hasn't had the "falling" dream? – as shown by Dahl et al. 2013. Apparently, humans aren't born with wariness of heights; instead, it emerges in association with the development of visual proprioception, usually when babies experience self-produced locomotion. Further, human infants tend to fear spiders and snakes, which may be an evolutionary retention (Hoehl et al. 2017), and many people are startled when things go bump in the night. What would you expect? After all, these responses helped our predecessors survive and, thus, are part of the reason we are here today.
32 Anderson 2018, p. 5.
33 Rosfort 2013.
34 Machado and Silva 2003.
35 Fischer et al. 2016.
36 Fischer et al. 2016. The authors show that making inferences about the physical world is supported by a network in the frontal and parietal lobes of the brain. One can think of babies' sequential development of individual skills related to physical perception and prediction as each one "coming online" when its neurological connections are established within this network during the first year of life (and likely beyond).
37 Herrmann et al. 2007, p. 1365. Similarly, Eckert et al. 2018 (p. 100) found that chimpanzees have intuitive statistical abilities for comparing absolute quantities that are on par with those of human infants, "suggesting that they constitute an evolutionary ancient ability."

38 Povinelli 2000.
39 Vonk 2020. See Vonk and Beran 2020 for a special issue of *Animal Behavior and Cognition* in honor of Povinelli's book from 2000.
40 Povinelli and Ballew 2012, p. 18.
41 Larsson and Falk 2025.
42 Most, if not all, of the apes had been captured in the jungles of Cameroon and transferred to Tenerife. Köhler noted that the classification of chimpanzees had not been clarified. Today, chimpanzees from Cameroon would probably be attributed to *Pan troglodytes troglodytes*.
43 This composition was not typical of chimpanzees in the wild, and quite likely the lack of adult males made it easier for Köhler to conduct his research. Although, as Boesch 2020 correctly points out, one must be careful about generalizing to all chimpanzees from such a small group, many of Köhler's discoveries have since been borne out by myriad studies on various groups of chimpanzees across Africa.
44 Goodall 1964, Pruetz and Bertolani 2007.
45 Köhler 1925, p. 277. Köhler believed that chimpanzees were very intelligent but not nearly as smart as humans because their lack of speech limited important components of thought.
46 Köhler 1925, p. 213.
47 After concluding that chimpanzees, indeed, sometimes showed insight, Köhler noted that an insightful "solution never appears in a disorder of blind impulses. It is one continuous, smooth action, which can be resolved into its parts *only by the imagination* of the onlooker; in *reality* they do *not* appear independently" (1925, pp. 198 and 200; italics his). Köhler also observed that the chimpanzees' insightful solutions tended to occur after a period of quiet perplexity or after a serendipitous discovery, and that when they finally happened it was usually all of a sudden. (Today, this is sometimes called a "flash of insight.")

As described by Machado and Silva 2003, the Russian physiologist Ivan Pavlov, who is famous for his research on classical conditioning (salivating dogs and dinner buzzers), *hated* Köhler's conclusion that apes sometimes arrived at solutions through insight rather than by trial and error. But then, Pavlov definitely had an (ahem) dog in the race. Köhler's ideas are, to this day, convincing because, as noted in this chapter, many of his discoveries anticipated modern (re)discoveries, his experiments seemed to be extremely well thought out, and his descriptions and conclusions seemed, for the most part, logical.
48 While testing for insight, Köhler discovered that caged chimpanzees generally had a terrible time figuring out that they could remove an obstacle such as a transport cage that was placed against the inside bars of their room, preventing them from reaching a nearby piece of fruit that was outside the room. An insightful solution to this situation finally occurred to the adult female, Tschego: "The 'obstacle' test was *not* solved ... by a series of imperceptible pushes involuntarily given to the cage in the act of stretching towards the prize. Quite the contrary: *during the lapse of two hours, Tschego did not move the cage one millimetre from its original position*, and when the solution arrived, the cage was not *shouldered* to one side, but *suddenly gripped with both hands*, and thrust back. It was a *genuine* solution" (Köhler 1925, p. 65).

Köhler thought the chimpanzees' difficulty in coming up with the (to humans) obvious solution of removing the obstructing box was because they perceived items that were merely touching as actually being connected. In other words, to the chimpanzees, the box seemed inseparable from the bars.

As with other discoveries, Köhler's findings here anticipated those of modern researchers who have (re)discovered that chimpanzees have difficulty interpreting flat images (Albiach-Serrano et al. 2015; Povinelli 2000; Seed et al. 2012). Albiach-Serrano et al. showed that by the time they are four years old, human children (but not apes) begin to develop an ability to grasp the inter-relationships of parts portrayed on two-dimensional (flat) surfaces. I think that this is likely due to development of the parts of the cerebral cortex that facilitate an ability to read, which is something else that happens in children of about this age. Chimpanzees do not read in the human sense, of course, even if they can learn to press a symbol to obtain a reward. Reading and its neurological substrates (which are largely understood) evolved very recently in human evolution. Köhler's perceptions about chimpanzees' difficulties in visually perceiving forms and shapes such as wound ropes or coiled wire (p. 235) or the separateness of items that have "optical firmness" because they are touching (p. 114) led him to conclude that "there must exist a kind of visual firmness, which makes the separating as an act of intelligence as difficult as the strongest nails would the actual pulling off of the board" (p. 114). Köhler was careful to note, however, that chimpanzees' vision is in other ways excellent. For example, "a chimpanzee, before making a wide jump at a considerable height, looks carefully to and fro across the intervening space. As an arboreal animal with immense range of spring and the need to use it, he *must* be able to measure distances" (p. 48).

49 Köhler 1925, p. 202.
50 Köhler 1925, p. 154.
51 Köhler 1925, p. 77.
52 Köhler 1925, p. 156; see also p. 164.
53 Köhler 1925, p. 157.
54 It also depends on an unconscious awareness of the relative positions of various parts of the body (called proprioception), which incorporates information from receptors in muscles, tendons and joints. These senses are, of course, also important for people, but do not seem to be as well developed in some respects (except, perhaps, in elite gymnasts). However, the maturation of proprioception in human infants is extremely important for the development of social skills, including the ability to empathize (Falk and Schofield 2018, see p. 53).
55 Rather than speculating about whether apes contemplate the underlying physics of their tool use, it seems eminently more reasonable to consider how they use and manipulate tools to achieve goals in their particular environments. This approach, called ecological psychology (more formally, perception/action theory), has recently and convincingly been applied to the discussion of folk physics in apes by Dorothy Fragaszy and Madhur Mangalam who theorize that animals perceive opportunities in their environments (such as a hanging piece of fruit), grasp an object (e.g., a stick), and apply it mechanically to (hopefully) acquire the perceived item: "Solving a problem by tooling depends upon the user's perception of affordances(s) for that individual to act with a given object to produce the intended mechanical effect on the target (i.e., to achieve their intended goal)" (Fragaszy and Mangalam 2020,

p. 466). The authors note that their theory is "embodied" because it does not specify what's going on in the animals' heads.
56 On the other hand, Köhler noted that, unlike humans, chimpanzees were not particularly good at acquiring out-of-reach fruits that required the use of tools that could not be seen at the same time as the treat (out of sight, out of mind), although Sultan managed to remember, retrieve, and solve a problem with a box that was in a different room from the coveted fruit (Köhler 1925, p. 54).
57 Köhler 1925, p. 277.
58 Osvath and Martin-Ordas 2014. This article offers a good review of current research on future-oriented cognition in apes.
59 Boesch and Boesch 1984.
60 Sanz and Morgan 2009.
61 Osvath and Martin-Ordas 2014. Apparently, the chimpanzee got better at ambushing his human visitors over time.
62 The parts of the frontal lobe of the brain that help you remember the reason why you walked into a room and then do whatever it is you had intended (known as "executive functions") are not nearly as large or well developed in great apes, nor are the nearby areas that are activated when you think about and plan for the future.
63 Köhler 1925, p. 324.
64 Anderson, 2012, 2013.

Chapter 2. Baskets Go to Ground

1 Falk 2011. Paleopolitics is as alive and well today as it was when the first *Homo erectus* specimen was discovered by Eugene Dubois in 1891 (Dubois 1894) and when Raymond Dart (Dart 1925) caused an uproar when he announced the discovery of a startling early hominin and assigned it new genus and species names (*Australopithecus africanus*).
2 Oh, how great it would be to have hands and feet that have the dexterity and precise control of human hands, as "quadrumanous" orangutans do!
3 A perceptive reader will have noticed that, while walking, both feet are briefly on the ground at the same time, which means that the stance phases of the two feet overlap a bit.
4 Napier 1967, quotation from p. 56.
5 Zihlman 2000. Get your crayons! This famous coloring book is a wonderful way to learn about (among other things) the comparative anatomy of apes, fossils, and humans.
6 Adolph, Hoch, and Cole 2018; Bründl et al. 2021; Falk 2009; Teulier, Lee, and Ulrich 2015; Lindsay 2019.
7 Gottlieb and DeLoache 2016.
8 Gupta 2019.
9 Adolph, Hoch, and Cole 2018, quotation from p. 703.
10 Adolph, Hoch, and Cole 2018, quotation from p. 707.
11 Adolph, Hoch and Cole 2018; Bründl et al. 2021; Plooij 1984.
12 Bründl et al. 2021, see table 3.

13 Except for crawling, the data for all milestones are from Bründl et al. 2021 (see also Plooij 1984). Data for crawling, which only humans do, are from Adolph, Burger, and Leo 2011.
14 Adolph, Burger, and Leo 2011, quotation from p. 2.
15 This topic might be worthy of a PhD dissertation! It wasn't just that legs got longer as hominin infants developed; the relative masses of their limbs and body proportions also changed in ways that affected the anatomy of mature individuals (Zihlman and Underwood 2019).
16 DeSilva et al. 2019.
17 Clarke 2019.
18 Heaton et al. 2019, Carlson et al. 2021.
19 Carlson et al. 2021, quotation from abstract. Further, Heaton et al. (2019) point out that the anatomy of the first foot bones Clarke discovered in a storage box prompted him to suggest that Little Foot's aligned big toe still had some degree of mobility that would have assisted in arboreality.
20 DeSilva et al. 2019. The famous footprints from Laetoli, Tanzania, that are dated to 3.66 MYA, and attributed by many to *Australopithecus afarensis*, are often viewed as corroborating evidence for a bipedal gait in that species.
21 Heaton et al. 2019.
22 Although a relatively complete skeleton, Lucy herself had little recovered in the way of foot bones.
23 Kappelman et al. 2016, quotation from p. 506.
24 DeSilva et al. 2019, quotation from p. 130.
25 DeSilva et al. 2019.
26 The Botanic Age described in this book should not be confused with the timespan that Roland Ennos' calls the Age of Wood in his book of the same name (2020). Ennos' Age of Wood overlaps with millions of years that are traditionally recognized as comprising the Stone Age (as they are in this book), an age that began around 3.5 million years ago on the heels of what is recognized here as the Botanic Age. It is not clear when, exactly, Ennos' Age of Wood began, although he suggests that its roots likely trace back at least as far as the early hominins that shared contemporary apes' ability to make botanical tools and that (much more recently) Neolithic people developed "methods people had used to weave branches together since before we came down from the trees" (Ennos, p. 95). While our books agree on many points, the present book focuses largely, but not entirely, on the first half of hominin evolution, whereas Ennos' book focuses largely, but not entirely, on the latter half. *The Botanic Age* does not focus much on *Homo sapiens*, but the reader can learn about the properties of wood and the huge impact that it had on the evolution of modern humans from Ennos' book. For example, he describes the construction of European warships during the early 1500s, which he calls "the High Wood Age" (Ennos, pp. 169–70). Ennos also explains that the Age of Wood ended at different times in different places. Thus, Britain's switch from burning wood to burning coal propelled it "and eventually the rest of the world, into an industrialized and urban society, and move[d] it out of the Age of Wood" (Ennos, p. 188), whereas "America, more than any other industrial nation, stayed rooted within the Age of Wood right up to the start of the twentieth century" (Ennos, p. 203). More of Ennos' ideas are discussed in chapter 7.

27 Many texts place the split between hominins and chimpanzees at ~5–7 MYA. This book uses 6.5 MYA, per timetree.org.
28 Couvreur et al. 2021, Bonnefille 2010, Cerling et al. 2011, Faith et al. 2018.
29 Joordens et al. 2019.
30 Joordens et al. 2019, Couvreur et al. 2021.
31 Couvreur et al. 2021, quotation from p. 29.
32 This is a modern-day version of the "savanna hypothesis" suggested by Raymond Dart when he announced the discovery of the first australopithecine in 1925.
33 Joordens et al. 2019.
34 Scerri et al. 2022. On p. 1, the authors point out that it is a mistake to think that ancestral hominins abandoned tropical forests altogether – i.e., that such forests were "relatively pristine environments left free from human influence – habitats deemed too hostile for humans throughout much of prehistory." In other words, although early hominins eventually broadened their horizons to occupy diverse habitats, some of them likely stayed put.
35 The situation is less clear for southern Africa, but see Macho et al. 2020.
36 Faith et al. 2018. The authors analyzed the fossil record of eastern African herbivores spanning the entire 7 million years of hominin existence by considering records for climatic and environmental change. They convincingly show that animal communities change (evolve) in response to environmental changes including those that affect vegetation.
37 As Domínguez-Rodrigo 2014 details, this definition of savanna updates the earlier idea that savannas were simply open grasslands: "A transition from dense forest to more open wooded areas to explain the emergence of hominins can potentially still preserve part of the spirit of the old 'savanna hypothesis' … but with a more gradual transition from one environment to another" (p. 69).
38 Fruth and Hohmann 1996.
39 Marlowe 2006.
40 Hicks 2010 , Koops et al. 2012, Tagg et al. 2013.
41 Koops et al. 2007.
42 Samson and Hunt 2012, quotation from p. 812.
43 Brownlow et al. 2001; Fruth, Tagg and Stewart 2018.
44 Samson and Hunt 2012.
45 Lindshield et al. 2021.
46 Lindshield et al. 2021.
47 Koops et al. 2007.
48 Fruth, Tagg, and Stewart 2018.
49 Bonnefille 2010 documents a second event (2.7–2.5 MYA) in which there was a shift from forests to drier conditions with more savannas, which corresponded with maximum expansion of glaciers in the northern hemisphere. In other words, there was a marked trend toward more open habitats during the first part of the Stone Age.
50 Ashbury et al. 2015.
51 Russon et al. 2007.
52 MacKinnon 1974.
53 Kuze et al. 2011.
54 Goodall 1971.

Chapter 3. Did You Make Your Nest This Morning?

1 Shultz et al. 2004.
2 Berger and Clarke 1995; Brain 1969. Eagles also preyed on other primates such as vervet monkeys, some of which have developed special alarm calls in response to seeing them.
3 Noticeable differences between the sexes in one or more features such as size of the teeth, body mass, or coloration is known as sexual dimorphism.
4 Coss 2021; Koops et al. 2007.
5 Such "mate guarding" is well known among various primates.
6 McNutt, Zipfel, and DeSilva 2018. The estimate of 1.5 MYA is tentative and based on the observation by McNutt et al. (p. 213) that "footprints from Ileret, Kenya demonstrate that human-like arches and a modern toe-off mechanism were present in *Homo* by 1.5 Ma."
7 Garcia-Martinez et al. 2021; Larson 2009; Larson 2013; Young et al. 2015.
8 Use of the term "Westernized" here follows Yetish and McGregor's (2019, p. 317) definition of referring "generally to life in a postindustrial economy with modern technology and focuses less on geographic location or specific cultural history." Such societies have also been called WEIRD – western, educated, industrialized, rich, and democratic (Henrich, Heine, and Norenzayan 2010).
9 Crittenden et al. 2018.
10 Shultz et al. 2004.
11 Fire-burned stone tools and bone fragments dated to 1.5 MYA in Kenya are the earliest indication that hominins (*Homo erectus*) were, at least, able to control fire. According to Chazan (2017), the archaeological record suggests that, although hominins that lived before 1 MYA may have opportunistically collected naturally occurring fire as a seasonal resource, the ability to maintain fires probably did not happen until between 400,000 and 200,000 years ago in association with the development of base camp sites. The ability to make fires would have emerged even more recently. If Chazan is right, terrestrially sleeping hominins lacked the means to use fire to counter the severe pressure from nocturnal predators for at least a million years. The adaptations in the duration and patterns of sleep that lessened the exposure of groups of hominins to nocturnal predators thus occurred over an extended period of time.
12 Samson 2021.
13 Watson and Buzsáki 2015; Le Bon 2020. Slow-wave sleep is abbreviated in the literature as SWS, or non-REM sleep, or NREM sleep.
14 Lacaux et al. 2021, quotation from p. 1.
15 Patel, Reddy, and Araujo 2024.
16 Samson and Nunn 2015.
17 Foster 2020. This publication reviews the daily circadian rhythms and the neuroanatomical underpinnings that control the sleep/wake cycles.
18 Blumberg and Rattenborg 2017; Fultz et al. 2019; Grubb and Lauritzen 2019. The brain creates a memory by, first, reactivating the neural activity (in the hippocampus of the temporal lobe) that accompanied the experience, which usually happens during SWS. The new memory is then transferred to the neocortex during REM sleep, where it is consolidated by the formation of connections (synapses) (Diekelmann and Born 2010).

19 Blumberg and Rattenborg 2017. In infants, twitching during REM provides sensory feedback that contributes to the development of sensorimotor neuronal networks, whereas SWS is associated with the pruning of synapses during brain development (Knoop, de Groot, and Dudink 2021).
20 Kalmbach, Booth, and Behn 2017, quotation from p. 1. Actually, studies of small-scale and Westernized societies as well as experimental evidence suggest that biphasic sleep (one longish sleep at night and an additional nap during the day) may, in fact, be the ancestral norm for humans (Samson 201).
21 Knoop, de Groot, and Dudink 2021.
22 Knowland et al. 2022. More specifically, SWS seems to occur when infants learn how to recognize speech sounds and to remember words, while REM sleep is associated more with infants' acquisition of the grammatical rules for making words and stringing them together into phrases and sentences (Frost and Monaghan 2017).
23 Yetish and McGregor 2019 (p. 323) observe that most populations with and without electricity get approximately 6.25–7.25 hours sleep per night, which "may reflect a species-typical predisposition."
24 Aeschbach and Borbély 1993; Nunn and Samson 2018.
25 Yetish and McGregor 2019.
26 Havercamp, Morimura, and Hirata 2021, quotation from p. 90. For more details, see Anderson et al. 2019.
27 Anderson et al. 2019; Freemon, McNew, and Adey 1971; Reinhardt 2020.
28 Chimpanzees experience 7–9 non-REM/REM cycles during the night, considerably more than the average of 4–5 cycles for people in (at least) Westernized societies (Freemon, McNew, and Adey 1971; Foster 2020), despite the fact that the total amount of REM sleep in chimpanzees is around 1.6 hours (Lesku et al. 2006, see table A1), which is very similar to the 1.6 to 1.7 hours, on average, for humans (Nunn and Samson 2018; Aeschbach and Borbély 1993). But see Panjwani et al. 2007 for lower total REM sleep of about 1.2 hours per night in healthy male soldiers from India. Although the total amount of REM sleep per night thus remained relatively steady during hominin evolution, the duration of the individual bouts of REM increased in humans as they became incorporated within a decreased number of sleep cycles.
29 Yetish and McGregor 2019 prefer to call traditional or preindustrial societies small-scale subsistence societies, or 4S for short.
30 Three 4S societies studied by Yetish and McGregor are the Tsimané from Bolivia, South America, and, from Africa, the Kalahari San from Namibia and the Hadza from Tanzania.
31 Crittenden et al. 2018.
32 Worthman and Melby 2002.
33 Worthman and Melby 2002, quotation from p. 79.
34 See McNamara and Bulkeley 2015 for review as well as discussion of the worldwide connection between dreaming and religious experiences.
35 Worthman and Melby 2002, quotation from p. 90.
36 Samson et al. 2017.
37 See Yetish and McGregor 2019, figure 21.3, for photographs of huts and sleeping spaces in other 4S societies.
38 Wadley et al. 2020.

39 Yetish and McGregor 2019, see table 21.1.
40 Samson 2021. Samson's reference to "the human sleep paradox" should not be confused with "paradoxical sleep," which is sometimes used to refer to REM sleep.
41 Samson 2021, quotation from p. 10. Samson more formally calls his idea the "social sleep hypothesis."
42 Samson 2021.
43 As one source charmingly puts it, "no matter how important sleep may be, it was adaptively deferred when the mountain lion entered the cave" (quoted from Spielman and Glovinsky 1991 by Perogamvros et al. 2020, p. 8).
44 Anderson et al. 2019, quotation from p. 11.
45 Falk 2000.
46 Coolidge and Wynn 2018; Wagner et al. 2004.
47 Simor et al. 2020. REM sleep is characterized by "phasic" periods in which individuals are disconnected from the external environment and, thus, especially vulnerable to threats, interspersed with "tonic" periods in which individuals become more alert and potentially capable of responding to external events. Curiously, sleep paralysis seems to occur during the tonic phases of REM, when people are more in touch with the external world. As explained by Simor et al. (2020, p. 5), "Environmental alertness during tonic periods may require the suppression of motor and visual imagery in order to efficiently anticipate, process and react to external stimulation." Thus, "from an evolutionary point of view ... tonic REM periods interspersed between more vulnerable phasic states might counteract the risks of environmental disconnection. Tonic periods might thus provide a transient alerting mechanism and reinstate external processing after periods of sensory disconnection" (Simor et al. 2020, quotation from p. 8).
48 Wagner et al. 2004.
49 Coolidge and Wynn 2018.
50 Schredl 2019.
51 da Cruz Ribeiro e Rodrigues 2019.
52 Dresler et al. 2014.
53 This is when I sometimes think something to myself like, "oh I'm dreaming, I don't want to wake up, I want to see how this turns out." As Dresler et al. detail, lucid dreaming is associated with activation of parts of the brain that function during "higher-order" thinking in awake people who are making future plans, remembering their pasts, or are conscious of being conscious. This goes beyond activation of the neurological networks associated with the perceptions and emotions that occur during non-lucid dreaming.
54 Lacaux et al. 2021, quotation from p. 1.
55 Coss 2021, quotation from p. 4. Recall from the previous chapter that two famous female australopithecines, Lucy and Little Foot, appear to have fallen to their deaths from trees, although the roles that possible predators may have played is not known.
56 Coss 2021, quotation from p. 7.
57 However, for girls, the most favored answers of "below" were only slightly higher than those for "side." Upon reflection, Coss attributed the relatively similar number of answers for "below" and "side" for girls to the fact that the beds of hominin females eventually became located on the ground, whereas this shift occurred earlier

in males who, thus, experienced greatly reduced vulnerability to arboreal predators compared to females during the Botanic Age.
58 Interestingly, it is not unusual for parents in various cultures to tell children that a ghost or monster will get them if they do not behave. Of course, this has a potential lasting negative impact on the severity of nighttime fears experienced by these children when they become adults (Coss and Blozis 2021). Who doesn't remember this refrain from James Whitcomb Riley's poem "Little Orphant Annie," which was published with the original title "The Elf Child" in 1885:
An' the Gobble-uns 'll git you
Ef you
Don't
Watch
Out!
59 Boyden, Pott, and Starks 2018, quotation from p. 101.
60 McNamara and Bulkeley 2015.
61 McNamara and Bulkeley 2015, quotation from p. 3.
62 Revonsuo 2000.
63 Chakraborty 2019.
64 Gilbert and Wilson 2007, quotation from p. 1352.
65 Gilbert and Wilson 2007, quotation from p. 1354.
66 So do other parts of the default system, although some become disconnected from each other, resulting in a loss of the consciousness that typifies wakefulness (Horovitz et al. 2009); see also Lacaux et al. 2021.
67 Coolidge and Wynn 2018.
68 Couvreur et al. 2021.
69 Nunn and Samson 2018; Samson and Nunn 2015.
70 McNamara and Bulkeley 2015.
71 Tanner and Zihlman 1976.

Chapter 4. From Tree Nests to Baby Carriers

1 de Castro et al. 2020; DeSilva et al. 2018; Larson 2013.
2 Young et al. 2015, quotation from p. 11829.
3 Ross 2001; Wall-Scheffler and Myers 2017.
4 Amaral 2008.
5 Plooij 1984, quotation from pp. 45–6.
6 Amaral 2008.
7 Tanner and Zihlman 1976, quotation from p. 600.
8 Fruth and Hohmann 1996.
9 Gunz et al. 2020. The estimated age of two years and five months for the Dikika infant is less than the original estimate of about three years.
10 DeSilva et al. 2018, quotation from p. 4.
11 See also Tanner and Zihlman 1976, p. 600.
12 Berecz et al. 2020. I used to think *Homo erectus* probably invented the first baby slings (Falk 2009). However, given that the Dikika (*Australopithecus afarensis*) infant's feet were somewhat adapted for walking, that arm morphology of *Australopithecus*

afarensis shows mothers likely climbed trees to make nests, and that stone tools had been invented by the time of *Australopithecus afarensis*, I think baby slings may have been invented by that species (and/or perhaps even some other earlier australopithecine).

13 Harmand et al. 2015.
14 McPherron et al. 2010.
15 Yetish and McGregor 2019.
16 See Konner (2005 and 2016) for recent updates about numerous hunter-gatherer groups including the !Kung, Hadza, Efe, and Aka from Africa, and the Aché from Paraguay and Agta from the Philippines.
17 Whiting 1981, quotation from p. 155.
18 Konner 1976.
19 Whiting 1981, quotation from p. 157. In a more recent study, Kushnick (2021) compared ethnographic information about whether specific cultures did or did not use baby carrying devices and, if so, whether they were slings, baskets, or rigid cradleboards (Barry and Paxson 1971) with information about the extent to which women in corresponding societies contributed to subsistence activities (Whyte 1978). The data for his study were from 77 societies selected from a database known as the Standard Cross-Cultural Sample, which contains a wide variety of information about 186 societies that are globally distributed (Murdock and White 1969). Somewhat surprisingly, a little over a quarter of the 77 cultures did not use baby carrying devices. Instead, infants were simply held in the arms or carried on a hip, usually facing the mother (based on my interpretation of data in Barry and Paxson 1971, p. 474, table 1), and were either clothed, in blankets, or had skin-to-skin contact with their carriers (Barry and Paxson 1971). Kushnik concluded that societies were more likely to have baby carriers when women played a bigger role in subsistence activities and that, "When they do, infant-carrying technology may arise to fulfil the dual needs of bringing the baby along to wherever that work must be done, and keeping the woman's hands free to engage in said work" (Kushnick 2021, quotation from p. 7).
20 John Barker kindly provided this description.
21 Whiting 1981, quotation from p. 161.
22 Whiting 1981, quotation from p. 175.
23 Whiting 1981, quotation from p. 161.
24 Konner 2005.
25 Ardelean et al. 2020. Although this date is much earlier than the date that many researchers have previously estimated people first arrived in the New World (i.e., around 15,000 years ago), it is nonetheless quite recent compared to the occupation dates of other parts of the world.
26 For example, people from southern Siberia are widely believed to have traveled over the Bering Land Bridge to enter North America.
27 Greenwald 2017.
28 Whiting 1981.
29 Greenwald 2017.
30 Greenwald 2017, quotation from p. 44.
31 Wall-Scheffler and Myers 2017.
32 Whiting 1981.

33 Konner 2005, quotation from p. 34.
34 Tanner and Zihlman 1976, quotation from p. 600.
35 Langley and Suddendorf 2020.
36 Langley and Suddendorf 2020.
37 McGrew 2013. However, McGrew notes (p. 4) that wild chimpanzees sometimes fold leaves to make a container for drinking water and that captive chimpanzees readily use containers supplied by humans. The closest captive chimpanzees get to using humanlike containers is to sometimes fill a coconut shell with water, which they carry elsewhere to drink. For all practical purposes, wild apes do not use carrying devices as humans do or their hominin predecessors did.
38 Marlowe 2006.
39 Wall-Scheffler and Myers 2017.
40 Róheim 1933, quotation from p. 255.

Chapter 5. First Came Wood, Then Came Stone

1 Berecz et al. 2020. For information about these slates, see Don Hitchcock's amazing website: https://donsmaps.com/gonnersdorf.html.
2 Anderson 2013, quotation from p. 27.
3 Anderson 2012.
4 Joordens et al. 2015, quotation from p. 228.
5 Beaumont and Vogel 2006, quotation from p. 222.
6 Anderson's ideas partly inspired what I call the "woven world hypothesis." See Falk 2024.
7 Hardy et al. 2020, quotation from p. 1.
8 Kvavadze et al. 2009.
9 Kehoe 2001, quotations from pp. 3 and 4.
10 Čufar et al. 2010.
11 Harmand et al. 2015. The authors propose the name "Lomekwian" for the tools, which predate similar-looking Oldowan tools from Olduvai Gorge in Tanzania by ~700,000 years.
12 Plummer et al. 2023; see also Braun et al. 2019.
13 McPherron et al. 2010. Although some disagree that the marks in question were inflicted by stone tools, I find this paper convincing.
14 Coolidge and Wynn 2016.
15 Coolidge and Wynn 2018.
16 To their credit, Coolidge and Wynn have been at the forefront in research on the evolution of sleep in early hominins (see chapter 3).
17 Coolidge and Wynn 2016, quotation from p. 387.
18 Steven Mithen (2024), a pioneer in the field of cognitive archaeology, offers a more thorough and nuanced analysis of what might be gleaned about the evolution of language from the material culture of various hominin species.
19 Coolidge and Wynn 2018, see pp. 113–14 for discussion.
20 Domínguez-Rodrigo et al. 2001. Although this paper is from 2001, the more recent analysis of Diez-Martín, Sánchez-Yustos, and de Luque (2018, quotation from p. 148) confirms that "the combination of strength and very precise morpho-functional

areas envision a number of ... activities, among which woodworking ... should still be taken into consideration."
21 Domínguez-Rodrigo et al. 2001.
22 Domínguez-Rodrigo et al. noted that tools that looked like the simpler earlier Oldowan tools were also found at Peninj and that these may have been used for different kinds of activities. The smaller Oldowan-like tools may, thus, have been associated more with processing animal remains and less challenging botanical matter than involved in working hardwoods like *Acacia*. This suggestion is consistent with the fact that tool traditions are accretionary through time (i.e., once the Acheulean emerged, the use of simpler stone tools still continued).
23 Lemorini et al. 2014, quotation from p. 10.
24 Gürbüz and Lycett 2021, quotation from p. 8. This paper reviews the evidence that shows that woodworking was important during the Stone Age.
25 As noted by Roland Ennos (2020, pp. 60–1), "The tools that modern great apes create are made for the moment and used immediately ... The same could be said for the stone knives and scrapers used by early humans to process animal carcasses. They were made and used on the spot – so the fact that early humans made stone tools ... hardly shows a step change in intelligence. The increasing success of humans is best explained by their development of their wooden tools, particularly weapons. These developments can also be used to chart the evolution of human intelligence."
26 Stibbard-Hawkes 2024.
27 Stibbard-Hawkes 2024, quotation from p. 8, emphasis his.
28 Parker and Gibson 1977.
29 Hohmann and Fruth 2007; Samuni, Wegdell, and Surbeck 2020; Surbeck and Hohmann 2008; Wakefield et al. 2019.
30 Wood and Gilby 2017.
31 Bugir, Butynski, and Hayward 2021. Looking across 13 studies covering four sites where chimpanzees hunt, Bugir qualifies that red colobus are clearly preferred when one takes into account the prey that are taken in excess of their abundance. If one simply counts prey captured because they are available, then infant and juvenile bushpigs are preferred (i.e., partly because they are abundant) more than red colobus.
32 Goodall 1986, see p. 282 for details.
33 Bugir, Butynski, and Hayward 2021. The upper limit for chimpanzee prey seems to be around 17 lb.
34 Wood and Gilby 2017.
35 Goodall 1986, see p. 306 for details.
36 But see p. 306 of Goodall 1986, in which she recounts an observation of a female chimpanzee chasing a colobus mother-infant pair through the canopy and seizing and beginning to eat the baby, all the while supporting her own two-week-old infant with one hand!
37 Goodall 1986; Surbeck and Hohmann 2008; Wood and Gilby 2017 (see p. 348).
38 Goodall 1986, quotation from p. 290.
39 According to Goodall (1986, pp. 550–2), chimpanzees have good aim and adult male chimpanzees engage in significantly more aimed throwing of sticks and stones than adult females.
40 In 1991 at the Mahale Mountains National Park in Tanzania, a female chimpanzee was observed using a modified tree branch to rouse a squirrel from a tree hole, which she then captured, killed, and ate (Huffman and Kalunde 1993). In 1995,

another Mahale female was observed poking a stick into a tree hollow, from which she removed a dying or already dead squirrel (Nakamura and Itoh 2008). Nearly a decade later, a male chimpanzee in the presence of other interested males attempted unsuccessfully to use a stick to obtain what was presumed to be a yellow-spotted rock hyrax from a nest in a constricted area underneath a large rock (Nakamura and Itoh 2008).

41 Domínguez-Rodrigo 2014.
42 Pruetz and Bertolani 2007.
43 Pruetz and Bertolani 2007.
44 Pruetz et al. 2015.
45 Pruetz et al. 2015.
46 Pruetz and Bertolani 2007.
47 Pruetz et al. 2015.
48 Pruetz et al. 2015, quotation from p. 2.
49 See Gumert, Low, and Malaivijitnond 2011 for review. Gruber, Clay, and Zuberbühler 2010 also reported that female long-tailed macaque monkeys used stone tools and smaller tools while feeding more often than males, and that females used stones to chip open sessile oysters more than males, whereas the latter used bigger rocks to pound open unattached foods more often than females.
50 Hernandez-Aguilar et al. 2007.
51 Domínguez-Rodrigo 2014.
52 Wood and Gilby 2017, quotation from p. 371.
53 Leder et al. 2024.
54 Schoch et al. 2015.
55 Barham et al. 2023; Belitzky, Goren-Inbar, and Werker 1991; Goren-Inbar, Werker, and Feibel 2002.
56 Barham et al. 2023.
57 Barham et al. 2023, quotation from p. 111.
58 Milks 2023, quotation from p. 35. See also Milks et al. 2022 for a glossary of wood from the entire archaeological record.
59 Berndt 1974, quotation from p. 72.
60 Allington-Jones 2015.
61 Allington-Jones 2015.
62 Wood and Gilby 2017.
63 Brightman 1996.
64 Brightman 1996, quotation from p. 705.
65 Brightman 1996.
66 It is also worth noting that females seem to be the inventive sex among the famous (and sexually dimorphic) Japanese macaque monkeys, as I have discussed elsewhere (Falk 2009, pp. 111–13).
67 Wall-Scheffler and Myers 2017; Whiting 1981.

Chapter 6. Babies Fall, Language Rises

1 Savage-Rumbaugh and Fields 2011. As I do, the authors believe that the loss of apelike clinging in hominin infants was at the heart of the eventual evolution of humanlike language. However, they also think that human language is distinct from ape communication because it is associated with a bifurcation of human consciousness

into two roles – I versus Me, representing doer and observer. Another way to look at the I/Me dichotomy is that it is a linguistic reflection of the fact that our nervous systems are fundamentally divided into sensory and motor domains. I perform (do) movements, but not the sensations that happen to me, which I instead perceive/observe.
2 Falk 2004.
3 Falk 2004, 2009.
4 Mothers would have been primarily responsible for their infants' survival, just as great ape mothers are, because, unlike today, early in our prehistory there would have been no conscious awareness of paternity.
5 Falk 2004, quotation from p. 503.
6 I now think my timing for the appearance of baby slings was deeply flawed because I believed, incorrectly as it turns out, that hominin infants would not have become helpless enough to require being carried in slings until brain size had enlarged to the point where it created an "obstetrical dilemma" in terms of feasible deliveries through maternal bony pelves that were constricted because of anatomical changes associated with the evolution of bipedalism. At that point (i.e., in relatively big-brained *Homo erectus*) babies had to be born sooner rather than later, that is, while their heads could still make the passage (as it were). They were, thus, born comparatively unfinished and not as far along in development as their ape cousins or earlier hominin predecessors. This book now shows that, regardless of *Homo erectus*' obstetrical problems, there is good reason to believe much earlier hominin infants lost their ability to cling to their mothers.
7 Falk 2009, quotation from p. 69.
8 Berecz et al. 2020.
9 Whiting 1981.
10 Koops et al. 2012.
11 Lindsay 2019 provides a convincing and welcome critique of the PTBD theory's failure to incorporate details pertaining to babies' evolution during the Botanic Age.
12 See, e.g., Plooij 1984 and chapter 2 for details.
13 Falk and Schofield 2018. I see becoming "late bloomers" as the earliest of three broad evo-devo trends that, together, altered infants in profound ways that charted the subsequent course of hominin evolution.
14 Savage-Rumbaugh and Fields 2011, quotation from p. 33.
15 Savage-Rumbaugh and Fields 2011.
16 Futagi, Toribe, and Suzuki 2012.
17 Lijowska et al. 1997; Thach and Lijowska 1996.
18 Lijowska et al. 1997. Some call the startle part of the sigh-startle reflex the Moro reflex, although others think that the curling in (adduction) of the limbs that immediately follows their abduction is separate from the Moro reflex (Futagi, Toribe, and Suzuki 2012). Interestingly, Rousseau et al. 2017 suggest that the Moro reflex (i.e., limbs flung out, then back in) is a phylogenetic retention that represents a nonverbal communication in which infants indicate they want to be picked up: "In human newborns who cannot move nor support their own weight by clinging to their mothers, the physiological behavior of grasping was transformed during evolution into a nonverbal communication behavior" (pp. 173–4) and "appeasement after being held

in the arms ... was significantly more effective than mother-talk alone to calm a crying newborn" (pp. 174–5).
19 See Lijowska et al. 1997 for details.
20 Thach and Lijowska 1996.
21 Lindsay 2019, quotation from p. 71.
22 Futagi, Toribe, and Suzuki 2012.
23 Barbu-Roth et al. 2015.
24 Lindsay 2019.
25 Newman 2004.
26 Wermke, Robb, and Schluter 2021.
27 Oller et al. 2013, 2019a, 2019b.
28 Melody (or tune) refers to the tones that are strung together in a rhythmic pattern that has varying pitches.
29 Melodic complexity entails an increase in the number of arc-like components of a melody (melody arcs) that can be assembled into multiple-arc melodies (Wermke, Robb, and Schluter 2021).
30 Bourvis et al. 2018; Falk 2009.
31 Levitt and Utman 1992.
32 Significantly, infants use the left sides of their brains to produce babbles (rather than the right side, which is favored for their other prosodic vocalizations discussed above), as adults do when they speak (Imada et al. 2006).
33 Ota, Davies-Jenkins, and Skarabela 2018, quotation from p. 1975.
34 Ota, Davies-Jenkins, and Skarabela 2018.
35 Interestingly, it is not unusual for adults to address lovers with baby talk words of endearment that were used to address them by their first loves – namely, their parents (e.g., sweetie-pie, cheeky-chops, etc.).
36 Ota, Davies-Jenkins, and Skarabela 2018, quotation from p. 1976.
37 So one shouldn't hesitate to talk baby talk to their infant – the more babies hear, the sooner and better they learn language (Falk 2009)!
38 Leongómez, Havlíček, and Roberts 2022.
39 Others also emphasize the primary role of breathing in the emergence of babbles and other aspects of language evolution; see Pouw and Fuchs 2021.
40 Descent of the larynx was associated with increased space above it in the neck (the supralaryngeal space) that is important for forming speech sounds.
41 Belyk and Brown 2017. In modern babies, "the first universal step towards prosody (intonation and rhythm), and hence language development for the young infant requires coordination of respiratory and laryngeal activity for the production of melody variations" (Wermke, Robb, and Schluter 2021, quotation from p. 2). Baby's first cries are reflexive, often coinciding with their first breaths. Robust cries are welcome in the delivery room as signs that babies' lungs are up-and-running (on land, as it were). Humans usually take their first gasps of air within about 10 seconds of their birth. The initial sharp inhale draws air into their lungs, causing them to expand, while the first exhale expels some of the mucous and amniotic fluid that accumulated prenatally. Thereafter, melodic vocalizations are produced when the vocal cords housed within the larynx are vibrated by air expired from the lungs, and the pitch of these vocalizations is modulated by stretching and relaxing the cords (Belyk

and Brown 2017). Infants do not develop an ability to assemble various sounds into words and sentences, etc., however, until their nervous systems mature enough for them to begin controlling the movements of not only their voice boxes, but also their supralaryngeal throats, mouths, and tongues.
42 Nishimura 2018.
43 Aboitiz 2018.
44 Belyk and Brown 2017.
45 As infants acquire knowledge of words and how to put them together, the left sides of their brains kick in to do the job. However, the prosodic components of language are controlled mostly by the right sides of their brains (Lindell 2013; Sammler et al. 2015). It is well known that most adults also have left-hemisphere dominance for the symbolic aspects of speech, grammar, etc., and right-hemisphere dominance for emotional/prosodic processing including tone of voice during speech. Apes are also neurologically asymmetrical to some degree, but not nearly to the same extent as people. Vast neurological networks that support language continue to be established on both sides of infants' brains as they continue to mature (Hertrich, Dietrich, and Ackermann 2020). Eventually, these networks contribute to the development of other advanced abilities such as playing the violin, mathematical skills, and reading. As this book argues, the origins of the novel behaviors and their neurological underpinnings that eventually enabled such cognitive advances likely trace back to the coping mechanisms of Botanic Age mothers and their increasingly helpless infants in response to disruptive effects from bipedalism (Falk 2009; Falk and Schofield 2018; see Ovando-Tellez et al. 2022 for more on creativity and neurological networks).
46 Filippi 2016.
47 Filippi 2016.
48 Kotz, Ravignani, and Fitch 2018.
49 Although the ability to keep a beat to music is unique to the human primate, it also evolved independently in other animals including some birds and sea mammals. For fun, watch this video of Snowball, the famous cockatoo: Eos Wetenschap, "A Compilation of the 14 Cockatoo Dance Moves," July 18, 2019, YouTube video, https://www.youtube.com/watch?v=YLM2KEBre3M.
50 Such entrainment is at the heart of dance – alone or in groups. Dance appears to be universal and is thought to promote social bonding. See Kotz, Ravignani, and Fitch 2018 for discussion.
51 Larsson, Ravignani, and Richter 2019. For a rare and fascinating example of chimpanzees engaging in synchronized bipedal walking, see Lameira, Eerola, and Ravignani 2019.
52 Larsson, Ravignani, and Richter 2019.
53 Kotz, Ravignani, and Fitch 2018.
54 Larsson and Falk 2025. Because this article is still in press, no page numbers are available for the quotations.

Chapter 7. What's Hobbit Got to Do with It?

1 Brown et al. 2004; Morwood et al. 2004.
2 Roberts et al. 2023; Sutikna et al. 2016.

3 Rizal et al. 2020; Roberts et al. 2023.
4 Certain details of LB1's teeth and skull resembled those of *Homo erectus* from nearby Java. Others such as short stature, pelvis, and remarkably short legs looked similar to *Australopithecus* (or *Homo habilis*) from Africa. The wrist and size of the brain were apelike, while the form of certain impressions stamped by the brain inside LB1's skull looked similar to those of living humans. Other parts of LB1's body were simply unique, such as shrugged forward shoulders and really long, flat feet. See Falk 2011 for details.
5 Jungers et al. 2009.
6 Baab 2016.
7 Argue 2022. Interestingly, at least one anthropologist speculates that some *Homo floresiensis* individuals might still be alive and hiding on the island of Flores (Forth 2022), though this is certainly a long shot!
8 Falk 2011.
9 Westaway et al. 2015, quotation from p. E604.
10 However, part of the reason for her tiny size may have been because her species had adapted to living on an island where resources like food were limited. This has been associated with dramatic miniaturization (called insular dwarfism) in certain animals, including primates (Bromham and Cardillo 2007).
11 Falk 2011.
12 Detroit et al. 2019.
13 *Homo floresiensis* and *Homo luzonensis* weren't the only surprising little and small-brained hominins that existed relatively recently. Another small-bodied, small-brained species, *Homo naledi*, came to light in 2013 in the Rising Star cave system in South Africa. Dated to 335,000 to ~240,000 years ago, these 15 or more individuals are controversial for a variety of reasons (Petraglia et al. 2023; Callaway 2023). Although *Homo naledi* is interesting in its own right, it is beyond the scope of this chapter's discussion about how small hominins may have gotten to islands in Southeast Asia.
14 Tocheri 2019.
15 Brumm et al. 2010.
16 Van den Bergh et al. 2016.
17 Dennell et al. 2014; Ruxton and Wilkinson 2012.
18 Kaifu 2017.
19 Dennell et al. 2014; Louys and Roberts 2020.
20 Morwood and Jungers 2009.
21 Morwood and Jungers 2009, quotation from p. 645.
22 Leppard 2015, quotation from p. 593. Such "rafting" may not be as far-fetched as it might seem. Numerous anecdotal rescues of people who floated for days or weeks on vegetative rafts "clearly illustrate the possibility of [hominin] individuals being washed up on remote islands following tsunamis and similar events" (Ruxton and Wilkinson 2012, quotation from p. 510).
23 Dennell et al. 2014, quotation from p. 16.
24 Westaway 2019.
25 Raven 2019; Roubik 2005.
26 Marivaux et al. 2023.

27 Ruxton and Wilkinson 2012.
28 This, indeed, may have been the case on Flores where the 700,000-year-old jaw and teeth are similar to, but even smaller than, those of LB1 and another individual in her species who lived some 600,000 years more recently (Van den Bergh et al. 2016).
29 Sivakumar 2010.
30 Raven 2019.
31 Chen and Lin 2018.
32 Raven 2019.
33 Hinojosa, Rivadeneira, and Thiel 2011.
34 Evans et al. 2020.
35 Roubik 2005.
36 Lin et al. 2011.
37 Nowack et al. 2015.
38 Takemoto 2004.
39 Anonymous, Atlanta Zoo 2020.
40 Chimpanzees from some sites build nests with greater structural support in windier conditions, and they tend to build thicker nests in cooler conditions and to nest higher in tree crowns during wet seasons (Stewart et al. 2018). Similarly, orangutans construct nests high in the canopy (Prasetyo et al. 2009), including in palm trees (Ancrenaz et al. 2015, Plate 1), and orangutans in a large peat-swamp forest in Borneo build nests in trees with features that protect from swaying in the wind and rain (Cheyne et al. 2013). They also build elaborate nests that may include special features made from botanic materials such as pillows, blankets, second bunks, and roofs of "braided branches, woven together to make a solid, nearly waterproof" cover (Prasetyo et al. 2009, quotation from p. 271).
41 A rare eyewitness description of an orangutan building a night nest seems particularly relevant: "It started to rain just as he was getting comfortable, and Mardianto ... pulled leaves over his body to blanket himself. I thought this was the end of it, until he flipped some more leaves over his head and disappeared! Suddenly, the only way to see Mardianto was with binoculars, and even then, all you could make out was a pair of eyes peering out from beneath a pile of leaves! Like a hermit in a hut, Mardianto was safe, happy and dry" (Borneo Orangutan Survival Australia 2022).
42 Raven 2019.
43 Baab 2016; Tocheri 2019.
44 The anatomy of both *Homo floresiensis* and *Homo luzonensis* indicates they likely had the flexibility to exploit both arboreal and terrestrial habitats (Baab 2016), which is thought to provide an advantage for primates recovering from cyclones (Zhang et al. 2019). It is also well known that relatively small-bodied species have an energetic advantage over larger-bodied hominins when establishing populations on islands with limited resources (Dennell et al. 2014).
45 Hinojosa, Rivadeneira, and Thiel 2011.
46 Ennos 2020. The earliest known ships are dated to less than 5,000 years ago.
47 Ennos 2020, quotations from pp. 95 and 96. He also notes that "not only Europeans ... used wickerwork. It is a technology that is found all around the world, suggesting that like the ax and the adze it was independently invented many times and widely shared" (pp. 96–7). Recall from chapter 4 that the same suggestion has been made for the invention of baby slings.

48 Ennos 2020, see p. 109.
49 Kealy, Louys, and O'Connor 2018.
50 Bird et al. 2019. Using analytical modeling, the authors explored how Australia and surrounding areas may have been colonized and concluded modern humans had the ability to plan and make open-sea voyages that lasted several days by at least 50,000 years ago.
51 Braje et al. 2020.
52 Pigati et al. 2023.
53 Hoffecker, Elias, and O'Rourke 2014. The authors accept the "Beringian standstill hypothesis" that posits the eventual colonizers of the Americas spent protracted time before 15,000 years ago in refugia on Beringia, only after which they pulled up stakes and dispersed south.
54 Braje et al. 2020, quotation from p. 14. The authors define the coastal route "as largely along shorelines, estuaries, littoral zones, river deltas, and coastal plains, where maritime peoples relied on coastal resources as the central focus of their subsistence" (p. 3).
55 Praetorius et al. 2023.
56 For much more detailed descriptions and discussion of wood inventions throughout the ages, see Ennos 2020.
57 Renn 2019. It is likely that reading and writing were invented in parallel by several civilizations, although Mesopotamia usually gets the credit.
58 Jain and Gupta 2021.
59 Bloom 2017; Bloom defines paper as "a mat of cellulose fibers that have been beaten in the presence of water, collected on a screen, and dried" (p. 52).
60 Dried and treated animal skins in the form of parchment were also favored writing material in Europe and Asia before paper was invented (Jain and Gupta 2021).
61 Bloom 2017, quotation from p. 51. The author also observes that "the historical geography of paper and papermaking concerns far more than the mere history of a material ... [it] encouraged a shift from oral to written culture and the development of various systems of notation, whether of language, mathematics, commercial transactions, music, or drawing and architectural drafting, quite apart from the invention and dissemination of printed books and images" (p. 64).
62 In addition to the rise of reading and writing a mere 5,000 years ago, other intellectual revolutions that affected hominin communication stemmed from the invention of spoken language (of course) hundreds of thousands, if not millions, of years ago and from information processing/communicating via the public Internet, which debuted only about 30 years ago (discussed in Falk and Schofield 2018). Stay tuned – things are speeding up when it comes to information technology!

Conclusion

1 Lacy and Ocobock 2024.
2 Falk and Schofield 2018.
3 In his splendid book, *The Age of Wood* (2020, p. 267), Roland Ennos theorizes that a related "fault line in history probably occurred around 1600, when we started to supplement the energy we obtained from firewood and charcoal – and gradually replaced them – with fossil fuels. This process stimulated urbanization, the birth of

science, and the rise of capitalism and industrialization." In the conclusion of his book (p. 279), Ennos wistfully hopes that "the human race" might one day "return to the more gentle delights of the Age of Wood."

4 One scholar defines myth as "a fictional or illusory product of the imagination, although it can be taken as real, literate, material, significatory, and/or suggestive of ... reality" (Mills 2020, p. 12). Another author comments on the "widely different opinions regarding the validity of oral traditions as factual" (Forth 2005, p. 16). In other words, one should not rule out the possibility that popular accounts might include kernels of truth.

5 The idea that the initial impetus for the evolution of advanced hominin cognition took place in trees has parallels, of course, in allegorical stories about Adam and Eve in the Garden of Eden, which appear in the creation stories of Judaism, Christianity, and Islam (the Abrahamic religions). The "tree of wisdom" in Buddhism refers to the tree that the Buddha was seated beneath when He received enlightenment. For details, see Liya 2018.

Not all of the relevant symbolism is so uplifting, however. Trees, after all, have been associated with undesirable falls – from grace and otherwise. One of the best-known examples is, of course, from Genesis 3 of the Old Testament: Eve is tempted by a talking snake to eat forbidden fruit from the tree of knowledge of good and evil and gives some to Adam, who also eats it. In what Christianity calls "the fall," God then curses all three. At a more down-to-earth level, this well-known (and variously interpreted) Mother Goose lullaby from the 1700s gets to the very heart of a deep-seated (perhaps even primordial) fear of infants falling from trees – see Grover and Richardson 1915 for a version that uses "Hush-a-bye" in place of "Rock-a-by." (In my opinion, if Mother Goose addressed a topic, it struck/strikes a chord!)

Rock-a-bye baby
on the treetop.
When the wind blows
the cradle will rock.
When the bough breaks,
the cradle will fall.
And down will come Baby,
Cradle and all.

6 Variations of the story of Noah's Ark appear in the Abrahamic religions. As told in Genesis 6:14–16, God instructed Noah to build an ark of wood, in which he, his family, and representatives of all living animals would be saved from an upcoming flood. Specifications for the ark included that it be pitched, its precise measurements, and details for a window and a door. After the flood, God instructed Noah and his sons to "be fruitful, and multiply, and replenish the earth" (Genesis 9:1). Again, at whatever level one takes it, this widely disseminated account entails the construction of a botanic vehicle (boat) for selectively rescuing and geographically dispersing life.

A different "ark" that entails basketry, saving a baby, and watercraft appears in Exodus 2 of the Old Testament. In order to save her son's life, baby Moses' mother waterproofed a basket of bullrushes with pitch and tar, put him in it, and placed the basket among reeds by the bank of the Nile, where it was discovered (by the Pharaoh's daughter) and he was saved. Again, this story is widespread. Similar narratives have been recorded for Sargon the Great (who became King of the Akkadian Empire, Mesopotamia) and Karna (who became a hero warrior in the Mahabharata ancient Indian epic). For details, see Lantz 2013.

7 One of my favorite parts of the boat tour is when the guide pauses along the river to point out where parts of two Tarzan movies starring Johnnie Weismuller were filmed in the junglelike surroundings – *Tarzan's Secret Treasure* in 1941 and *Tarzan's New York Adventure* in 1942. What could be more thrilling than seeing Tarzan swinging on vines through the trees while bellowing his iconic call? Shades of the Botanic Age, indeed! For excitement, however, it might be rivaled by another film that was also made at Wakulla Springs (in 1954) – what child of the 1950s could ever forget *Creature from the Black Lagoon*.

8 Quoted from McCoy 2022.
9 Gomez-Misserian 2023.
10 See chapter 5 for details.

References

Aboitiz, Francisco. 2018. "A Brain for Speech. Evolutionary Continuity in Primate and Human Auditory-Vocal Processing." *Frontiers in Neuroscience* 12: 174. https://doi.org/10.3389/fnins.2018.00174.

Adolph, Karen E., Sarah E. Berger, and Andrew J. Leo. 2011. "Developmental Continuity? Crawling, Cruising, and Walking." *Developmental Science* 14 (2): 306–18. https://doi.org/10.1111/j.1467-7687.2010.00981.x.

Adolph, Karen E., Justine E. Hoch, and Whitney G. Cole. 2018. "Development (of Walking): 15 Suggestions." *Trends in Cognitive Sciences* 22 (8): 699–711. https://doi.org/10.1016/j.tics.2018.05.010.

Aeschbach, Daniel, and Alexander A. Borbély. 1993. "All-Night Dynamics of the Human Sleep EEG." *Journal of Sleep Research* 2 (2): 70–81. https://doi.org/10.1111/j.1365-2869.1993.tb00065.x.

Albiach-Serrano, Anna, Carla Sebastián-Enesco, Amanda Seed, Fernando Colmenares, and Josep Call. 2015. "Comparing Humans and Nonhuman Great Apes in the Broken Cloth Problem: Is Their Knowledge Causal or Perceptual?" *Journal of Experimental Child Psychology* 139: 174–89. https://doi.org/10.1016/j.jecp.2015.06.004.

Allington-Jones, Lu. 2015. "The Clacton Spear: The Last One Hundred Years." *Archaeological Journal* 172 (2): 273–96. https://doi.org/10.1080/00665983.2015.1008839.

Amaral, Lia Q. 2008. "Mechanical Analysis of Infant Carrying in Hominoids." *Naturwissenschaften* 95 (4): 281–92. https://doi.org/10.1007/s00114-007-0325-0.

Ancrenaz, Marc, Felicity Oram, Laurentius Ambu, Isabelle Lackman, Eddie Ahmad, Hamisah Elahan, Harjinder Kler, Nicola K. Abram, and Erik Meijaard. 2015. "Of *Pongo*, Palms and Perceptions: A Multidisciplinary Assessment of Bornean Orang-utans *Pongo pygmaeus* in an Oil Palm Context." *Oryx* 49 (3): 465–72. https://doi.org/10.1017/S0030605313001270.

Anderson, Helen. 2012. "Crossing the Line: The Early Expression of Pattern in Middle Stone Age Africa." *Journal of World Prehistory* 25 (3–4): 183–204. https://doi.org/10.1007/s10963-012-9061-2.

Anderson, Helen. 2013. "A Distinguishing Skill Art, Language, and Complex Cognition." *Journal of Consciousness Studies* 20 (3–4): 6–32.

Anderson, James R. 2018. "Chimpanzees and Death." *Philosophical Transactions of the Royal Society B: Biological Sciences* 373 (1754): 20170257. https://doi.org/10.1098/rstb.2017.0257.

Anderson, James R., Mabel Y.L. Ang, Louise C. Lock, and Iris Weiche. 2019. "Nesting, Sleeping, and Nighttime Behaviors in Wild and Captive Great Apes." *Primates*: 1–12. https://doi.org/10.1007/s10329-019-00723-2.

Ardelean, Ciprian F., Lorena Becerra-Valdivia, Mikkel Winther Pedersen, Jean-Luc Schwenninger, Charles G. Oviatt, Juan I. Macías-Quintero, Joaquin Arroyo-Cabrales, Martin Sikora, Yam Zul E. Ocampo-Díaz, and Igor I. Rubio-Cisneros. 2020. "Evidence of Human Occupation in Mexico around the Last Glacial Maximum." *Nature* 584 (7819): 87–92. https://doi.org/10.1038/s41586-020-2509-0.

Argue, Debbie. 2022. *Little Species, Big Mystery*. Melbourne: Melbourne University Press.

Ashbury, A.M., M.R. Posa, L.P. Dunkel, B. Spillmann, S.S. Atmoko, C.P. van Schaik, and M.A. van Noordwijk. 2015. "Why Do Orangutans Leave the Trees? Terrestrial Behavior among Wild Bornean Orangutans (*Pongo pygmaeus wurmbii*) at Tuanan, Central Kalimantan." *American Journal of Primatology* 77 (11): 1216–29. https://doi.org/10.1002/ajp.22460.

Atlanta Zoo. n.d. "Bornean Orangutan." Accessed May 11, 2022. https://zooatlanta.org/animal/bornean-orangutan/.

Baab, Karen L. 2016. "The Place of *Homo floresiensis* in Human Evolution." *Journal of Anthropological Sciences* 94: 5–18. https://doi.org/10.4436/jass.94024.

Barbu-Roth, Marianne, David I. Anderson, Ryan J. Streeter, Marie Combrouze, Juana Park, Brooke Schultz, Joseph J. Campos, François Goffinet, and Joëlle Provasi. 2015. "Why Does Infant Stepping Disappear and Can It Be Stimulated by Optic Flow?" *Child Development* 86 (2): 441–55. https://doi.org/10.1111/cdev.12305.

Barham, L., G.A.T. Duller, I. Candy, C. Scott, C.R. Cartwright, J.R. Peterson, C. Kabukcu, M.S. Chapot, F. Melia, V. Rots, N. George, N. Taipale, P. Gethin, and P. Nkombwe. 2023. "Evidence for the Earliest Structural Use of Wood at Least 476,000 Years Ago." *Nature* 622 (7981): 107–11. https://doi.org/10.1038/s41586-023-06557-9.

Barry, Herbert, and Leonora M. Paxson. 1971. "Infancy and Early Childhood: Cross-Cultural Codes 2." *Ethnology* 10 (4): 466–508. https://doi.org/10.2307/3773177.

Beaumont, Peter B., and John C. Vogel. 2006. "On a Timescale for the Past Million Years of Human History in Central South Africa." *South African Journal of Science* 102 (5): 217–28.

Belitzky, Shmuel, Naama Goren-Inbar, and Ella Werker. 1991. "A Middle Pleistocene Wooden Plank with Man-Made Polish." *Journal of Human Evolution* 20 (4): 349–53. https://doi.org/10.1016/0047-2484(91)90015-N.

Belyk, Michel, and Steven Brown. 2017. "The Origins of the Vocal Brain in Humans." *Neuroscience & Biobehavioral Reviews* 77: 177–93. https://doi.org/10.1016/j.neubiorev.2017.03.014.

Berecz, Bernadett, Mel Cyrille, Ulrika Casselbrant, Sarah Oleksak, and Henrik Norholt. 2020. "Carrying Human Infants – An Evolutionary Heritage." *Infant Behavior and Development* 60: 101460. https://doi.org/10.1016/j.infbeh.2020.101460.

Berger, Lee R., and Ron J. Clarke. 1995. "Eagle Involvement in Accumulation of the Taung Child Fauna." *Journal of Human Evolution* 29 (3): 275–99. https://doi.org/10.1006/jhev.1995.1060.

Berndt, Catherine H. 1974. "Digging Sticks and Spears, or the Two Sex Model." In *Women's Role in Aboriginal Society*, edited by F. Gale, 64–80. Canberra: Australian Institute of Aboriginal Studies.

Bird, Michael I., Scott A. Condie, Sue O'Connor, Damien O'Grady, Christian Reepmeyer, Sean Ulm, Mojca Zega, Frédérik Saltré, and Corey J.A. Bradshaw. 2019. "Early Human Settlement of Sahul Was Not an Accident." *Scientific Reports* 9 (1): 8220. https://doi.org/10.1038/s41598-019-42946-9.

Bloom, Jonathan M. 2017. "Papermaking: The Historical Diffusion of an Ancient Technique." In *Mobilities of Knowledge*, edited by Heike Jöns, Peter Meusburger, and Michael Heffernan, 51–66. Cham, Switzerland: Springer.

Blumberg, Mark S., John A. Lesku, Paul-Antoine Libourel, Markus H. Schmidt, and Niels C. Rattenborg. 2020. "What Is REM Sleep?" *Current Biology* 30 (1): R38–R49. https://doi.org/10.1016/j.cub.2019.11.045.

Blumberg, M.S., and Niels Christian Rattenborg. 2017. "Decomposing the Evolution of Sleep: Comparative and Developmental Approaches." In *Evolution of Nervous Systems*, edited by Jon H. Kaas, 523–45. Cambridge, MA: Elsevier.

Boesch, Christophe. 2020. "Mothers, Environment, and Ontogeny Affect Cognition." *Animal Behavior and Cognition* 7 (3): 474–89. https://doi.org/10.26451/abc.07.03.13.2020.

Boesch, Christophe, and Hedwige Boesch. 1984. "Mental Map in Wild Chimpanzees: An Analysis of Hammer Transports for Nut Cracking." *Primates* 25 (2): 160–70. https://doi.org/10.1007/BF02382388.

Bonnefille, Raymonde. 2010. "Cenozoic Vegetation, Climate Changes and Hominid Evolution in Tropical Africa." *Global and Planetary Change* 72 (4): 390–411. https://doi.org/10.1016/j.gloplacha.2010.01.015.

Borneo Orangutan Survival Australia. n.d. "Everyone Takes Cover in the Rain." Accessed May 11, 2022. https://www.orangutans.com.au/news-bos-international/everyone-takes-cover-in-the-rain/.

Bourvis, Nadège, Magi Singer, Catherine Saint Georges, Nicolas Bodeau, Mohamed Chetouani, David Cohen, and Ruth Feldman. 2018. "Pre-linguistic Infants Employ Complex Communicative Loops to Engage Mothers in Social Exchanges and Repair Interaction Ruptures." *Royal Society Open Science* 5 (1): 170274. https://doi.org/10.1098/rsos.170274.

Boyden, Sean D., Martha Pott, and Philip T. Starks. 2018. "An Evolutionary Perspective on Night Terrors." *Evolution, Medicine, and Public Health* 2018 (1): 100–5. https://doi.org/10.1093/emph/eoy010.

Brain, C.K. 1969. "The Probable Role of Leopards as Predators of the Swartkrans Australopithecines." *South African Archaeological Bulletin* 24: 170–1. https://doi.org/10.2307/3888296.

Braje, Todd J., Jon M. Erlandson, Torben C. Rick, Loren Davis, Tom Dillehay, Daryl W. Fedje, Duane Froese, Amy Gusick, Quentin Mackie, and Duncan McLaren. 2020. "Fladmark+ 40: What Have We Learned about a Potential Pacific Coast Peopling of the Americas?" *American Antiquity* 85 (1): 1–21. https://doi.org/10.1017/aaq.2019.80.

Braun, David R., Vera Aldeias, Will Archer, J. Ramon Arrowsmith, Niguss Baraki, Christopher J. Campisano, Alan L. Deino, Erin N. DiMaggio, Guillaume Dupont-Nivet, and Blade Engda. 2019. "Earliest Known Oldowan Artifacts at >2.58 Ma from Ledi-Geraru, Ethiopia, Highlight Early Technological Diversity." *Proceedings of the National Academy of Sciences* 116 (24): 11712–17. https://doi.org/10.1073/pnas.1820177116.

Brightman, Robert. 1996. "The Sexual Division of Foraging Labor: Biology, Taboo, and Gender Politics." *Comparative Studies in Society and History* 38 (4): 687–729. https://doi.org/10.1017/S0010417500020508.

Bromham, L., and M. Cardillo. 2007. "Primates Follow the 'Island Rule': Implications for Interpreting *Homo floresiensis*." *Biology Letters* 3 (4): 398–400. https://doi.org/10.1098/rsbl.2007.0113.

Brown, P., T. Sutikna, M.J. Morwood, R.P. Soejono, Jatmiko, E.W. Saptomo, and R.A. Due. 2004. "A New Small-Bodied Hominin from the Late Pleistocene of Flores, Indonesia." *Nature* 431 (7012): 1055–61. https://doi.org/10.1038/nature02999.

Brownlow, A.R., A.J. Plumptre, V. Reynolds, and R. Ward. 2001. "Sources of Variation in the Nesting Behavior of Chimpanzees (*Pan troglodytes schweinfurthii*) in the Budongo Forest, Uganda." *American Journal of Primatology: Official Journal of the American Society of Primatologists* 55 (1): 49–55. https://doi.org/10.1002/ajp.1038.

Brumm, A., G.M. Jensen, G.D. van den Bergh, M.J. Morwood, I. Kurniawan, F. Aziz, and M. Storey. 2010. "Hominins on Flores, Indonesia, by One Million Years Ago." *Nature* 464 (7289): 748–52. https://doi.org/10.1038/nature08844.

Bründl, A.C., P.J. Tkaczynski, G. Nohon Kohou, C. Boesch, R.M. Wittig, and C. Crockford. 2021. "Systematic Mapping of Developmental Milestones in Wild Chimpanzees." *Developmental Science* 24 (1): e12988. https://doi.org/10.1111/desc.12988.

Brunet, M., F. Guy, D. Pilbeam, H.T. Mackaye, A. Likius, D. Ahounta, A. Beauvilain, C. Blondel, H. Bocherens, J.R. Boisserie, L. De Bonis, Y. Coppens, J. Dejax, C. Denys, P. Duringer, V. Eisenmann, G. Fanone, P. Fronty, D. Geraads, T. Lehmann, F. Lihoreau, A. Louchart, A. Mahamat, G. Merceron, G. Mouchelin, O. Otero, P. Pelaez Campomanes, M. Ponce De Leon, J.C. Rage, M. Sapanet, M. Schuster, J. Sudre, P. Tassy, X. Valentin, P. Vignaud, L. Viriot, A. Zazzo, and C. Zollikofer. 2002. "A New Hominid from the Upper Miocene of Chad, Central Africa." *Nature* 418 (6894): 145–51. https://doi.org/10.1038/nature00879.

Bugir, Cassandra K., Thomas M. Butynski, and Matt W. Hayward. 2021. "Prey Preferences of the Chimpanzee (*Pan troglodytes*)." *Ecology and Evolution* 11 (12): 7138–46. https://doi.org/10.1002/ece3.7633.

Buzsaki, Gyorgy. 2006. *Rhythms of the Brain*. New York: Oxford University Press.

Callaway, Ewen. 2023. "Sharp Criticism of Controversial Ancient-Human Claims Tests eLife's Revamped Peer-Review Model." *Nature* 620 (7972): 13–14. https://doi.org/10.1038/d41586-023-02415-w.

Carlson, Kristian J., David J. Green, Tea Jashashvili, Travis R. Pickering, Jason L. Heaton, Amélie Beaudet, Dominic Stratford, Robin Crompton, Kathleen

Kuman, and Laurent Bruxelles. 2021. "The Pectoral Girdle of StW 573 ('Little Foot') and Its Implications for Shoulder Evolution in the Hominina." *Journal of Human Evolution*: 102983. https://doi.org/10.1016/j.jhevol.2021.102983.

Cerling, T.E., J.G. Wynn, S.A. Andanje, M.I. Bird, D.K. Korir, N.E. Levin, W. Mace, A.N. Macharia, J. Quade, and C.H. Remien. 2011. "Woody Cover and Hominin Environments in the Past 6 Million Years." *Nature* 476 (7358): 51–6. https://doi.org/10.1038/nature10306.

Chakraborty, Roma. 2019. "Understanding Dreams from an Evolutionary Perspective: A Critical Study." *Philosophical Papers: Journal of the Department of Philosophy* 17: 49–59. http://ir.nbu.ac.in/handle/123456789/3466.

Chazan, Michael. 2017. "Toward a Long Prehistory of Fire." *Current Anthropology* 58 (S16): S351–S359. https://doi.org/10.1086/691988.

Chen, Tzu-Hsin, and Kuan-Hui Elaine Lin. 2018. "Distinguishing the Windthrow and Hydrogeological Effects of Typhoon Impact on Agricultural Lands: An Integrative OBIA and PPGIS Approach." *International Journal of Remote Sensing* 39 (1): 131–48. https://doi.org/10.1080/01431161.2017.1382741.

Cheyne, Susan M., Dominic Rowland, Andrea Höing, and Simon J. Husson. 2013. "How Orangutans Choose Where to Sleep: Comparison of Nest Site Variables." *Asian Primates Journal* 3 (1): 13–17.

Clarke, R.J. 2019. "Excavation, Reconstruction and Taphonomy of the StW 573 *Australopithecus prometheus* Skeleton from Sterkfontein Caves, South Africa." *Journal of Human Evolution* 127 (2): 41–53. https://doi.org/10.1016/j.jhevol.2018.11.010.

Coolidge, Frederick L., and Thomas Grant Wynn. 2016. "An Introduction to Cognitive Archaeology." *Current Directions in Psychological Science* 25 (6): 386–92. https://doi.org/10.1177/0963721416657085.

Coolidge, Frederick L., and Thomas Grant Wynn. 2018. *The Rise of Homo sapiens: The Evolution of Modern Thinking*. New York: Oxford University Press.

Coss, Richard G. 2021. "Something Scary Is Out There: Remembrances of Where the Threat Was Located by Preschool Children and Adults with Nighttime Fear." *Evolutionary Psychological Science*: 1–15. https://doi.org/10.1007/s40806-021-00279-9.

Coss, Richard G., and Shelley A. Blozis. 2021. "Something Scary Is Out There II: The Interplay of Childhood Experiences, Relict Sexual Dinichism, and Cross-Cultural Differences in Spatial Fears." *Evolutionary Psychological Science* 7 (4): 359–79. https://doi.org/10.1007/s40806-021-00289-7.

Couvreur, Thomas L.P., Gilles Dauby, Anne Blach-Overgaard, Vincent Deblauwe, Steven Dessein, Vincent Droissart, Oliver J. Hardy, David J. Harris, Steven B.

Janssens, and Alexandra C. Ley. 2021. "Tectonics, Climate and the Diversification of the Tropical African Terrestrial Flora and Fauna." *Biological Reviews* 96 (1): 16–51. https://doi.org/10.1111/brv.12644.

Crittenden, Alyssa N., David R. Samson, Kristen N. Herlosky, Ibrahim A. Mabulla, Audax Z.P. Mabulla, and James J. McKenna. 2018. "Infant Co-sleeping Patterns and Maternal Sleep Quality among Hadza Hunter-Gatherers." *Sleep Health* 4 (6): 527–34. https://doi.org/10.1016/j.sleh.2018.10.005.

Čufar, Katarina, Bernd Kromer, Tjaša Tolar, and Anton Velušček. 2010. "Dating of 4th Millennium BC Pile-Dwellings on Ljubljansko barje, Slovenia." *Journal of Archaeological Science* 37 (8): 2031–9. https://doi.org/10.1016/j.jas.2010.03.008.

da Cruz Ribeiro e Rodrigues, Christiano. 2019. "Giuseppe Tartini's 'Devil's Trill' Sonata: An Arrangement and Recording for Solo Violin." Doctoral dissertation, Arizona State University.

Dahl, Audun, Joseph J. Campos, David I. Anderson, Ichiro Uchiyama, David C. Witherington, Mika Ueno, Laure Poutrain-Lejeune, and Marianne Barbu-Roth. 2013. "The Epigenesis of Wariness of Heights." *Psychological Science* 24 (7): 1361–7. https://doi.org/10.1177/0956797613476047.

Dart, R.A. 1925. "The Taungs Skull." *Nature* 116: 462. https://doi.org/10.1038/116462a0.

Darwin, Charles. 1859. *On the Origin of Species by Means of Natural Selection.* London: J. Murray.

Darwin, Charles. 1871. *The Descent of Man, and Selection in Relation to Sex.* New York: D. Appleton and Company.

de Castro, José María Bermúdez, Marina Martínez de Pinillos, Lucía López-Polín, Laura Martín-Francés, Cecilia García-Campos, Mario Modesto-Mata, Jordi Rosell, and María Martinón-Torres. 2020. "A Descriptive and Comparative Study of Two Early Pleistocene Immature Scapulae from the TD6.2 Level of the Gran Dolina Cave Site (Sierra de Atapuerca, Spain)." *Journal of Human Evolution* 139: 102689. https://doi.org/10.1016/j.jhevol.2019.102689.

Dennell, Robin W., Julien Louys, Hannah J. O'Regan, and David M. Wilkinson. 2014. "The Origins and Persistence of *Homo floresiensis* on Flores: Biogeographical and Ecological Perspectives." *Quaternary Science Reviews* 96: 98–107. https://doi.org/10.1016/j.quascirev.2013.06.031.

DeSilva, J.M., C.M. Gill, T.C. Prang, M.A. Bredella, and Z. Alemseged. 2018. "A Nearly Complete Foot from Dikika, Ethiopia and Its Implications for the Ontogeny and Function of *Australopithecus afarensis*." *Science Advances* 4 (7): eaar7723. https://doi.org/10.1126/sciadv.aar7723.

DeSilva, Jeremy, Ellison McNutt, Julien Benoit, and Bernhard Zipfel. 2019. "One Small Step: A Review of Plio-Pleistocene Hominin Foot Evolution." *American Journal of Physical Anthropology* 168: 63–140. https://doi.org/10.1002/ajpa.23750.

Detroit, F., A.S. Mijares, J. Corny, G. Daver, C. Zanolli, E. Dizon, E. Robles, R. Grun, and P.J. Piper. 2019. "A New Species of *Homo* from the Late Pleistocene of the Philippines." *Nature* 568 (7751): 181–6. https://doi.org/10.1038/s41586-019-1067-9.

Diekelmann, Susanne, and Jan Born. 2010. "Slow-Wave Sleep Takes the Leading Role in Memory Reorganization." *Nature Reviews Neuroscience* 11 (3): 218. https://doi.org/10.1038/nrn2762-c2.

Diez-Martín, Fernando, Policarpo Sánchez-Yustos, and Luis de Luque. 2018. "The East African Early Acheulean of Peninj (Lake Natron, Tanzania)." In *The Emergence of the Acheulean in East Africa and Beyond*, edited by Rosalia Gallotti and Margherita Mussi, 129–51. Springer.

Domínguez-Rodrigo, Manuel. 2014. "Is the "Savanna Hypothesis" a Dead Concept for Explaining the Emergence of the Earliest Hominins?" *Current Anthropology* 55 (1): 59–81. https://doi.org/10.1086/674530.

Domínguez-Rodrigo, Manuel, Jordi Serrallonga, Jordi Juan-Tresserras, Luis Alcala, and Luis Luque. 2001. "Woodworking Activities by Early Humans: A Plant Residue Analysis on Acheulian Stone Tools from Peninj (Tanzania)." *Journal of Human Evolution* 40 (4): 289–99. https://doi.org/10.1006/jhev.2000.0466.

Dresler, Martin, Leandra Eibl, Christian F.J. Fischer, Renate Wehrle, Victor I. Spoormaker, Axel Steiger, Michael Czisch, and Marcel Pawlowski. 2014. "Volitional Components of Consciousness Vary across Wakefulness, Dreaming and Lucid Dreaming." *Frontiers in Psychology* 4: 987. https://doi.org/10.3389/fpsyg.2013.00987.

Dubois, Eugene. 1894. *Pithecanthropus erectus. Eine Menschenaehnliche Uebergangsform aus Java*. Batavia: Landsdrukkerij.

Eckert, J., J. Call, J. Hermes, E. Herrmann, and H. Rakoczy. 2018. "Intuitive Statistical Inferences in Chimpanzees and Humans Follow Weber's Law." *Cognition* 180: 99–107. https://doi.org/10.1016/j.cognition.2018.07.004.

Ennos, Roland. 2020. *The Age of Wood*. New York: Scribner.

Evans, B.J., M.T. Gansauge, M.W. Tocheri, M.A. Schillaci, T. Sutikna, Jatmiko, E.W. Saptomo, A. Klegarth, A.J. Tosi, D.J. Melnick, and M. Meyer. 2020. "Mitogenomics of Macaques (*Macaca*) across Wallace's Line in the Context of Modern Human Dispersals." *Journal of Human Evolution* 146 (i–vi): 102852. https://doi.org/10.1016/j.jhevol.2020.102852.

Faith, J. Tyler, John Rowan, Andrew Du, and Paul L. Koch. 2018. "Plio-Pleistocene Decline of African Megaherbivores: No Evidence for Ancient Hominin Impacts." *Science* 362 (6417): 938–41. https://doi.org/10.1126/science.aau2728.

Falk, Dean. 2000. *Primate Diversity*. New York: W.W. Norton.

Falk, Dean. 2004. "Prelinguistic Evolution in Early Hominins: Whence Motherese?" *Behavioral and Brain Sciences* 27 (4): 491–541. https://doi.org/10.1017/S0140525X04000111.

Falk, Dean. 2009. *Finding Our Tongues: Mothers, Infants and The Origins of Language*. New York: Basic Books.

Falk, Dean. 2011. *The Fossil Chronicles: How Two Controversial Discoveries Changed Our View of Human Evolution*. Berkeley: University of California Press.

Falk, Dean. 2024. "Evolution of the Mind's Eye: A Possible Explanation for Why Early Hominins Produced Geometric Figures before Representational Ones." In *A Quest for Understanding the Evolution of Human Constructs of Reality*, edited by Giriraj Kumar, 101–16. New Delhi: Research India Press.

Falk, Dean, and Eve Penelope Schofield. 2018. *Geeks, Genes, and the Evolution of Asperger Syndrome*. Albuquerque: University of New Mexico Press.

Filippi, Piera. 2016. "Emotional and Interactional Prosody across Animal Communication Systems: A Comparative Approach to the Emergence of Language." *Frontiers in Psychology* 7: 1393. https://doi.org/10.3389/fpsyg.2016.01393.

Fischer, J., J.G. Mikhael, J.B. Tenenbaum, and N. Kanwisher. 2016. "Functional Neuroanatomy of Intuitive Physical Inference." *Proceedings of the National Academy of Sciences* 113 (34): E5072–81. https://doi.org/10.1073/pnas.1610344113.

Forth, Gregory. 2005. "Hominids, Hairy Hominoids and the Science of Humanity." *Anthropology Today* 21 (3): 13–17.

Forth, Gregory. 2022. *Between Ape and Human: An Anthropologist on the Trail of a Hidden Hominoid*. New York: Pegasus Books.

Foster, Russell G. 2020. "Sleep, Circadian Rhythms and Health." *Interface Focus* 10 (3): 20190098. https://doi.org/10.1098/rsfs.2019.0098.

Fragaszy, D.M., and M. Mangalam. 2020. "Folks Physics in the Twenty-First Century: Understanding Tooling as Embodied." *Animal Behavior and Cognition* 7 (3): 457–73. https://doi.org/10.26451/abc.07.03.12.2020.

Freemon, Frank R., James J. McNew, and W. Ross Adey. 1971. "Chimpanzee Sleep Stages." *Electroencephalography and Clinical Neurophysiology* 31 (5): 485–9. https://doi.org/10.1016/0013-4694(71)90169-6.

Frost, Rebecca L.A., and Padraic Monaghan. 2017. "Sleep-Driven Computations in Speech Processing." *PloS ONE* 12 (1): e0169538. https://doi.org/10.1371/journal.pone.0169538.

Fruth, Barbara, and Gottfried Hohmann. 1994. "Nests: Living Artefacts of Recent Apes?" *Current Anthropology* 35 (3): 310–11. https://doi.org/10.1086/204281.

Fruth, Barbara, and Gottfried Hohmann. 1996. "Nest Building Behavior in the Great Apes: The Great Leap Forward?" In *Great Ape Societies*, edited by

L. Marchant, W. McGrew, and T. Nishida, 225–40. Cambridge, UK: Cambridge University Press.

Fruth, B., N. Tagg, and F. Stewart. 2018. "Sleep and Nesting Behavior in Primates: A Review." *American Journal of Biological Anthropology* 166 (3): 499–509. https://doi.org/10.1002/ajpa.23373.

Fultz, N.E., G. Bonmassar, K. Setsompop, R.A. Stickgold, B.R. Rosen, J.R. Polimeni, and L.D. Lewis. 2019. "Coupled Electrophysiological, Hemodynamic, and Cerebrospinal Fluid Oscillations in Human Sleep." *Science* 366 (6465): 628–31. https://doi.org/10.1126/science.aax5440.

Futagi, Yasuyuki, Yasuhisa Toribe, and Yasuhiro Suzuki. 2012. "The Grasp Reflex and Moro Reflex in Infants: Hierarchy of Primitive Reflex Responses." *International Journal of Pediatrics* 2012: 191562. https://doi.org/10.1155/2012/191562.

Garcia-Martinez, D., D.J. Green, and J.M. Bermudez de Castro. 2021. "Evolutionary Development of the Homo Antecessor Scapulae (Gran Dolina Site, Atapuerca) Suggests a Modern-Like Development for Lower Pleistocene *Homo*." *Scientific Reports* 11 (1): 4102. https://doi.org/10.1038/s41598-021-83039-w.

Gilbert, D.T., and T.D. Wilson. 2007. "Prospection: Experiencing the Future." *Science* 317 (5843): 1351–4. https://doi.org/10.1126/science.1144161.

Gomez-Misserian, Gebriela. 2023. "The Stickman's Last Bough." Accessed November 2, 2023. https://gardenandgun.com/articles/the-stickmans-last-bough/.

Goodall, Jane. 1964. "Tool-Using and Aimed Throwing in a Community of Free-Living Chimpanzees." *Nature* 201: 1264–6. https://doi.org/10.1038/2011264a0.

Goodall, Jane. 1971. *In the Shadow of Man*. Boston: Houghton Mifflin.

Goodall, Jane. 1986. *The Chimpanzees of Gombe: Patterns of Behavior*. Cambridge, MA: Belknap Press of Harvard University Press.

Goren-Inbar, Naama, Ella Werker, and Craig S. Feibel. 2002. *The Acheulian Site of Gesher Benot Ya'aqov, Israel: The Wood Assemblage*. Oxford: Oxbow.

Gottlieb, Alma, and Judy S. DeLoache. 2016. *A World of Babies: Imagined Childcare Guides for Eight Societies*. Cambridge: Cambridge University Press.

Grabowski, Mark, and William L. Jungers. 2017. "Evidence of a Chimpanzee-Sized Ancestor of Humans but a Gibbon-Sized Ancestor of Apes." *Nature Communications* 8 (1): 1–10. https://doi.org/10.1038/s41467-017-00997-4.

Greenwald, A. 2017. "Mediating Women's Time Allocation Trade-Offs: Basketry Cradle Technology in California and the Maintenance of Maternal Foraging Efficiency." *Journal of California and Great Basin Anthropology* 37 (1): 38–47. https://www.jstor.org/stable/45408196.

Grover, Eulalie Osgood, and Frederick Richardson. 1915. *Mother Goose*. Chicago: PF Volland.

Groves, C.P. 2018. "The Latest Thinking about the Taxonomy of Great Apes." *International Zoo Yearbook* 52 (1): 16–24. https://doi.org/10.1111/izy.12173.

Grubb, Søren, and Martin Lauritzen. 2019. "Deep Sleep Drives Brain Fluid Oscillations." *Science* 366 (6465): 572–3. https://doi.org/10.1126/science.aaz5191.

Gruber, Thibaud, Zanna Clay, and Klaus Zuberbühler. 2010. "A Comparison of Bonobo and Chimpanzee Tool Use: Evidence for a Female Bias in the *Pan* Lineage." *Animal Behaviour* 80 (6): 1023–33. https://doi.org/10.1016/j.anbehav.2010.09.005.

Gumert, Michael D., Low Kuan Hoong, and Suchinda Malaivijitnond. 2011. "Sex Differences in the Stone Tool-Use Behavior of a Wild Population of Burmese Long-Tailed Macaques (*Macaca fascicularis aurea*)." *American Journal of Primatology* 73 (12): 1239–49. https://doi.org/10.1002/ajp.20996.

Gunz, Philipp, Simon Neubauer, Dean Falk, Paul Tafforeau, Adeline Le Cabec, Tanya M. Smith, William H. Kimbel, Fred Spoor, and Zeresenay Alemseged. 2020. "*Australopithecus afarensis* Endocasts Suggest Ape-Like Brain Organization and Prolonged Brain Growth." *Science Advances* 6 (14): eaaz4729. https://doi.org/10.1126/sciadv.aaz4729.

Gupta, Sujata. 2019. "Culture Helps Shape When Babies Learn to Walk." *Science News*, September 10. https://www.sciencenews.org/article/culture-helps-shape-when-babies-learn-walk.

Gürbüz, Rebecca Biermann, and Stephen J. Lycett. 2021. "Could Woodworking Have Driven Lithic Tool Selection?" *Journal of Human Evolution* 156: 102999. https://doi.org/10.1016/j.jhevol.2021.102999.

Hardy, Bruce L., M.-H. Moncel, Céline Kerfant, Matthieu Lebon, Ludovic Bellot-Gurlet, and Nicolas Mélard. 2020. "Direct Evidence of Neanderthal Fibre Technology and Its Cognitive and Behavioral Implications." *Scientific Reports* 10 (1): 4889. https://doi.org/10.1038/s41598-020-61839-w.

Harmand, Sonia, Jason E. Lewis, Craig S. Feibel, Christopher J. Lepre, Sandrine Prat, Arnaud Lenoble, Xavier Boës, Rhonda L. Quinn, Michel Brenet, and Adrian Arroyo. 2015. "3.3-Million-Year-Old Stone Tools from Lomekwi 3, West Turkana, Kenya." *Nature* 521 (7552): 310–15. https://doi.org/10.1038/nature14464.

Havercamp, Kristin, Naruki Morimura, and Satoshi Hirata. 2021. "Sleep Patterns of Aging Chimpanzees (*Pan troglodytes*)." *International Journal of Primatology* 42 (1): 89–104. https://doi.org/10.1007/s10764-020-00190-3.

Heaton, Jason L., Travis Rayne Pickering, Kristian J. Carlson, Robin H. Crompton, Tea Jashashvili, Amélie Beaudet, Laurent Bruxelles, Kathleen Kuman, Andrea J. Heile, and Dominic Stratford. 2019. "The Long Limb Bones of the

StW 573 *Australopithecus* Skeleton from Sterkfontein Member 2: Descriptions and Proportions." *Journal of Human Evolution* 133: 167–97. https://doi.org/10.1016/j.jhevol.2019.05.015.

Henrich, Joseph, Steven J. Heine, and Ara Norenzayan. 2010. "The Weirdest People in the World?" *Behavioral and Brain Sciences* 33 (2–3): 61–83. https://doi.org/10.1017/S0140525X0999152X.

Hernandez-Aguilar, R. Adriana, Jim Moore, and Travis Rayne Pickering. 2007. "Savanna Chimpanzees Use Tools to Harvest the Underground Storage Organs of Plants." *Proceedings of the National Academy of Sciences* 104 (49): 19210–13. https://doi.org/10.1073/pnas.0707929104.

Herrmann, Esther, Josep Call, María Victoria Hernández-Lloreda, Brian Hare, and Michael Tomasello. 2007. "Humans Have Evolved Specialized Skills of Social Cognition: The Cultural Intelligence Hypothesis." *Science* 317 (5843): 1360–6. https://doi.org/10.1126/science.1146282.

Hertrich, Ingo, Susanne Dietrich, and Hermann Ackermann. 2020. "The Margins of the Language Network in the Brain." *Frontiers in Communication* 5: 519955. https://doi.org/10.3389/fcomm.2020.519955.

Hicks, Thurston Cleveland. 2010. "A Chimpanzee Mega-Culture? Exploring Behavioral Continuity in Pan troglodytes schweinfurthii across Northern DR Congo." PhD thesis, University of Amsterdam.

Hinojosa, Ivan A., Marcelo M. Rivadeneira, and Martin Thiel. 2011. "Temporal and Spatial Distribution of Floating Objects in Coastal Waters of Central–Southern Chile and Patagonian Fjords." *Continental Shelf Research* 31 (3–4): 172–86. https://doi.org/10.1016/j.csr.2010.04.013.

Hoehl, Stefanie, Kahl Hellmer, Maria Johansson, and Gustaf Gredebäck. 2017. "Itsy Bitsy Spider…: Infants React with Increased Arousal to Spiders and Snakes." *Frontiers in Psychology* 8: 1710. https://doi.org/10.3389/fpsyg.2017.01710.

Hoffecker, John F., Scott A. Elias, and Dennis H. O'Rourke. 2014. "Out of Beringia?" *Science* 343 (6174): 979–80. https://doi.org/10.1126/science.1250768.

Hohmann, Gottfried, and Barbara Fruth. 2007. "New Records on Prey Capture and Meat Eating by Bonobos at Lui Kotale, Salonga National Park, Democratic Republic of Congo." *Folia Primatologica* 79 (2): 103–10. https://doi.org/10.1159/000110679.

Horovitz, Silvina G., Allen R. Braun, Walter S. Carr, Dante Picchioni, Thomas J. Balkin, Masaki Fukunaga, and Jeff H. Duyn. 2009. "Decoupling of the Brain's Default Mode Network during Deep Sleep." *Proceedings of the National Academy of Sciences* 106 (27): 11376–81. https://doi.org/10.1073/pnas.0901435106.

Huffman, Michael A., and Mohamedi Seifu Kalunde. 1993. "Tool-Assisted Predation on a Squirrel by a Female Chimpanzee in the Mahale Mountains, Tanzania." *Primates* 34: 93–8. https://doi.org/10.1007/BF02381285.

Imada, T., Y. Zhang, M. Cheour, S. Taulu, A. Ahonen, and P.K. Kuhl. 2006. "Infant Speech Perception Activates Broca's Area: A Developmental Magnetoencephalography Study." *Neuroreport* 17 (10): 957–62. https://journals.lww.com/neuroreport/abstract/2006/07170/infant_speech_perception_activates_broca_s_area__a.3.aspx.

Jain, Prerna, and Charu Gupta. 2021. "A Sustainable Journey of Handmade Paper from Past to Present: A Review." *Problemy Ekorozwoju* 16 (2): 234–44. https://doi.org/10.35784/pe.2021.2.25.

Joordens, Josephine C.A., Francesco d'Errico, Frank P. Wesselingh, Stephen Munro, John De Vos, Jakob Wallinga, Christina Ankjærgaard, Tony Reimann, Jan R. Wijbrans, and Klaudia F. Kuiper. 2015. "*Homo erectus* at Trinil on Java Used Shells for Tool Production and Engraving." *Nature* 518 (7538): 228–31. https://doi.org/10.1038/nature13962.

Joordens, Josephine C.A., Craig S. Feibel, Hubert B. Vonhof, Anne S. Schulp, and Dick Kroon. 2019. "Relevance of the Eastern African Coastal Forest for Early Hominin Biogeography." *Journal of Human Evolution* 131: 176–202. https://doi.org/10.1016/j.jhevol.2019.03.012.

Jungers, William L., William E.H. Harcourt-Smith, R.E. Wunderlich, Matthew W. Tocheri, Susan G. Larson, T. Sutikna, Rhokus Awe Due, and Michael J. Morwood. 2009. "The Foot of *Homo floresiensis*." *Nature* 459 (7243): 81–4. https://doi.org/10.1038/nature07989.

Kaifu, Yousuke. 2017. "Archaic Hominin Populations in Asia Before the Arrival of Modern Humans: Their Phylogeny and Implications for the 'Southern Denisovans.'" *Current Anthropology* 58 (S17): S418–S433. https://doi.org/10.1086/694318.

Kalmbach, Kelsey, Victoria Booth, and C.G. Behn. 2017. "REM Sleep Complicates Period Adding Bifurcations from Monophasic to Polyphasic Sleep Behavior in a Sleep–Wake Regulatory Network Model for Human Sleep." *arXiv*:1710.05494. https://doi.org/10.48550/arXiv.1710.05494.

Kappelman, J., R.A. Ketcham, S. Pearce, L. Todd, W. Akins, M.W. Colbert, M. Feseha, J.A. Maisano, and A. Witzel. 2016. "Perimortem Fractures in Lucy Suggest Mortality from Fall Out of Tall Tree." *Nature* 537 (7621): 503–7. https://doi.org/10.1038/nature19332.

Kealy, Shimona, Julien Louys, and Sue O'Connor. 2018. "Least-Cost Pathway Models Indicate Northern Human Dispersal from Sunda to Sahul." *Journal of Human Evolution* 125: 59–70. https://doi.org/10.1016/j.jhevol.2018.10.003.

Kehoe, Alice B. 2001. "From Spirit Cave to the Blackfoot Rez: The Importance of Twined Fabric in North American Indian Societies." In *Fleeting Identities: Perishable Material Culture in Archaeological Research*, edited by P. Drooker, 210–25. Carbondale: Southern Illinois University Press.

Knoop, Marit S., Eline R. de Groot, and Jeroen Dudink. 2021. "Current Ideas about the Roles of Rapid Eye Movement and Non–Rapid Eye Movement Sleep in Brain Development." *Acta Paediatrica* 110 (1): 36–44. https://doi.org/10.1111/apa.15485.

Knowland, Victoria C.P., Sam Berens, M. Gareth Gaskell, Sarah A. Walker, and Lisa-Marie Henderson. 2022. "Does the Maturation of Early Sleep Patterns Predict Language Ability at School Entry? A Born in Bradford Study." *Journal of Child Language* 49 (1): 1–23. https://doi.org/10.1017/S0305000920000677.

Köhler, W. 1925. *The Mentality of Apes*. Translated by E. Winter. New York: Harcourt, Brace. (Original work published 1924.)

Konner, Melvin. 1976. "Maternal Care, Infant Behavior and Development among the !Kung." In *Kalahari Hunter-Gatherers: Studies of the !Kung San and Their Neighbors*, edited by R. Lee and I. Devore, 218–45. Cambridge, MA: Harvard University Press.

Konner, Melvin. 2005. "Hunter-Gatherer Infancy and Childhood: The !Kung and Others." In *Hunter-Gatherer Childhoods: Evolutionary, Developmental, and Cultural Perspectives*, edited by Barry S. Hewlett and Michael E. Lamb, 19–64. Piscataway, NJ: Aldine Transactions.

Konner, Melvin. 2016. "Hunter-Gatherer Infancy and Childhood in the Context of Human Evolution." In *Childhood: Origins, Evolution, and Implications*, edited by C.L. Meehan and A.N. Crittenden, 123–54. Santa Fe, NM: School for Advanced Research Press.

Koops, Kathelijne. 2011. "Chimpanzees in the Seringbara Region of the Nimba Mountains." In *The Chimpanzees of Bossou and Nimba*, edited by Tetsuro Matsuzawa, Tatyana Humle and Yukimaru Sugiyama, 277–87. Tokyo: Springer Japan. https://doi.org/10.1007/978-4-431-53921-6_29.

Koops, Kathelijne. 2020. "Chimpanzee Termite Fishing Etiquette." *Nature Human Behaviour* 4 (9): 887–8. https://doi.org/10.1038/s41562-020-0895-9.

Koops, Kathelijne, Tatyana Humle, Elisabeth H.M. Sterck, and Tetsuro Matsuzawa. 2007. "Ground-Nesting by the Chimpanzees of the Nimba Mountains, Guinea: Environmentally or Socially Determined?" *American Journal of Primatology: Official Journal of the American Society of Primatologists* 69 (4): 407–19. https://doi.org/10.1002/ajp.20358.

Koops, Kathelijne, William C. McGrew, Tetsuro Matsuzawa, and Leslie A. Knapp. 2012. "Terrestrial Nest-Building by Wild Chimpanzees (*Pan troglodytes*): Implications for the Tree-to-Ground Sleep Transition in Early Hominins." *American Journal of Physical Anthropology* 148 (3): 351–61. https://doi.org/10.1002/ajpa.22056.

Kotz, Sonja A., Andrea Ravignani, and William T. Fitch. 2018. "The Evolution of Rhythm Processing." *Trends in Cognitive Sciences* 22 (10): 896–910. https://doi.org/10.1016/j.tics.2018.08.002.

Kuhlwilm, M., M. de Manuel, A. Nater, M. P. Greminger, M. Krutzen, and T. Marques-Bonet. 2016. "Evolution and Demography of the Great Apes." *Current Opinion in Genetics and Development* 41: 124–9. https://doi.org/10.1016/j.gde.2016.09.005.

Kushnick, Geoff. 2021. "The Cradle of Humankind." In *The Oxford Handbook of Evolutionary Psychology and Parenting*, edited by V.A. Weekes-Shackelford and T.K. Shackelford, 115–34. New York: Oxford University Press.

Kuze, Noko, Hiroto Kawabata, Saika Yamazaki, Tomoko Kanamori, Titol Peter Malim, and Henry Bernard. 2011. "A Wild Borneo Orangutan Carries Large Numbers of Branches on the Neck for Feeding and Nest Building in the Danum Valley Conservation Area." *Primate Research* 27 (1): 21–6. https://doi.org/10.2354/psj.27.007.

Kvavadze, Eliso, Ofer Bar-Yosef, Anna Belfer-Cohen, Elisabetta Boaretto, Nino Jakeli, Zinovi Matskevich, and Tengiz Meshveliani. 2009. "30,000-Year-Old Wild Flax Fibers." *Science* 325 (5946): 1359. https://doi.org/10.1126/science.1175404.

Lacaux, C., T. Andrillon, C. Bastoul, Y. Idir, A. Fonteix-Galet, I. Arnulf, and D. Oudiette. 2021. "Sleep Onset Is a Creative Sweet Spot." *Science Advances* 7 (50): eabj5866. https://doi.org/10.1126/sciadv.abj5866.

Lacy, Sarah, and Cara Ocobock. 2024. "Woman the Hunter: The Archaeological Evidence." *American Anthropologist* 126 (1): 19–31. https://doi.org/10.1111/aman.13914.

Lameira, Adriano R., Tuomas Eerola, and Andrea Ravignani. 2019. "Coupled Whole-Body Rhythmic Entrainment Between Two Chimpanzees." *Scientific Reports* 9 (1): 18914. https://doi.org/10.1038/s41598-019-55360-y.

Langley, Michelle C., and Thomas Suddendorf. 2020. "Mobile Containers in Human Cognitive Evolution Studies: Understudied and Underrepresented." *Evolutionary Anthropology: Issues, News, and Reviews* 29 (6): 299–309. https://doi.org/10.1002/evan.21857.

Lantz, Sabio. 2013. "Basket Cases: Moses, Sargon, and Karna." *Triangulations* (blog). December 19. https://triangulations.wordpress.com/2013/12/19/basket-cases-moses-sargon-and-karna/.

Larson, Susan G. 2009. "Evolution of the Hominin Shoulder: Early *Homo*." In *The First Humans: Origin and Early Evolution of the Genus Homo*, edited by Frederick E. Grine, John G. Fleagle, and Richard E. Leakey, 65–75. New York: Springer.

Larson, Susan G. 2013. "Shoulder Morphology in Early Hominin Evolution." In *The Paleobiology of Australopithecus*, edited by Kaye E. Reed, John G. Fleagle, and Richard E. Leakey, 247–61. New York: Springer.

Larsson, Matz, and Dean Falk. 2025. "Direct Effects of Bipedalism on Early Hominin Fetuses Stimulated Later Musical and Linguistic Evolution." *Current Anthropology*. (Forum article to appear with international peer commentaries. Accepted for publication.)

Larsson, Matz, Joachim Richter, and Andrea Ravignani. 2019. "Bipedal Steps in the Development of Rhythmic Behavior in Humans." *Music & Science* 2: 1–14. https://doi.org/10.1177/2059204319892617.

Le Bon, Olivier. 2020. "Relationships between REM and NREM in the NREM-REM Sleep Cycle: A Review on Competing Concepts." *Sleep Medicine* 70: 6–16. https://doi.org/10.1016/j.sleep.2020.02.004.

Leder, Dirk, Jens Lehmann, Annemieke Milks, Tim Koddenberg, Michael Sietz, Matthias Vogel, Utz Böhner, and Thomas Terberger. 2024. "The Wooden Artifacts from Schöningen's Spear Horizon and Their Place in Human Evolution." *Proceedings of the National Academy of Sciences* 121 (15): e2320484121. https://doi.org/10.1073/pnas.2320484121.

Lemorini, Cristina, Thomas W. Plummer, David R. Braun, Alyssa N. Crittenden, Peter W. Ditchfield, Laura C. Bishop, Fritz Hertel, James S. Oliver, Frank W. Marlowe, and Margaret J. Schoeninger. 2014. "Old Stones' Song: Use-Wear Experiments and Analysis of the Oldowan Quartz and Quartzite Assemblage from Kanjera South (Kenya)." *Journal of Human Evolution* 72: 10–25. https://doi.org/10.1016/j.jhevol.2014.03.002.

Leongómez, Juan David, Jan Havlíček, and S. Craig Roberts. 2022. "Musicality in Human Vocal Communication: An Evolutionary Perspective." *Philosophical Transactions of the Royal Society B* 377 (1841): 20200391. https://doi.org/10.1098/rstb.2020.0391.

Leppard, Thomas P. 2015. "Passive Dispersal versus Strategic Dispersal in Island Colonization by Hominins." *Current Anthropology* 56 (4): 590–5. https://doi.org/10.1086/682325.

Lesku, John A., Timothy C. Roth II, Charles J. Amlaner, and Steven L. Lima. 2006. "A Phylogenetic Analysis of Sleep Architecture in Mammals: The Integration of Anatomy, Physiology, and Ecology." *The American Naturalist* 168 (4): 441–53. https://doi.org/10.1086/506973.

Levitt, A.G., and J.G. Utman. 1992. "From Babbling towards the Sound Systems of English and French: A Longitudinal Two-Case Study." *Journal of Child Language* 19 (1): 19–49. https://doi.org/10.1017/S0305000900013611.

Lijowska, Anna S., Nevada W. Reed, Barbara A. Mertins Chiodini, and Bradley T. Thach. 1997. "Sequential Arousal and Airway-Defensive Behavior of Infants in Asphyxial Sleep Environments." *Journal of Applied Physiology* 83 (1): 219–28. https://doi.org/10.1152/jappl.1997.83.1.219.

Lin, Teng-Chiu, Steven P. Hamburg, Kuo-Chuan Lin, Lih-Jih Wang, Chung-Te Chang, Yue-Joe Hsia, Matthew A. Vadeboncoeur, Cathy M. Mabry McMullen, and Chiung-Pin Liu. 2011. "Typhoon Disturbance and Forest Dynamics: Lessons from a Northwest Pacific Subtropical Forest." *Ecosystems* 14 (1): 127–43. https://doi.org/10.1007/s10021-010-9399-1.

Lindell, Annukka. 2013. "Continuities in Emotion Lateralization in Human and Non-Human Primates." *Frontiers in Human Neuroscience* 7. https://doi.org/10.3389/fnhum.2013.00464.

Lindsay, David. 2019. "When the Bough Breaks: A Contribution to Falk's Hypothesis." *Journal of Language Evolution* 4 (1): 71–7. https://doi.org/10.1093/jole/lzy011.

Lindshield, S., R.A. Hernandez-Aguilar, A.H. Korstjens, L.F. Marchant, V. Narat, P.I. Ndiaye, H. Ogawa, A.K. Piel, J.D. Pruetz, F.A. Stewart, K.L. van Leeuwen, E.G. Wessling, and M. Yoshikawa. 2021. "Chimpanzees (*Pan troglodytes*) in Savanna Landscapes." *Evolutionary Anthropology* 30 (6): 399–420. https://doi.org/10.1002/evan.21924.

Liya, Sally. 2018. "The Use of Trees as Symbols in the World Religions." *a ollada dasibilia* (blog), edited by Carlos Penela, July 14. https://aolladadasibila.wordpress.com/2018/07/14/the-use-of-trees-as-symbols-in-the-world-religions-by-sally-liya/.

Louys, Julien, and Patrick Roberts. 2020. "Environmental Drivers of Megafauna and Hominin Extinction in Southeast Asia." *Nature*: 1–5. https://doi.org/10.1038/s41586-020-2810-y.

Machado, Armando, and Francisco J. Silva. 2003. "You Can Lead an Ape to a Tool, but …: A Review of Povinelli's *Folk Physics for Apes: The Chimpanzee's Theory of How the World Works*." *Journal of the Experimental Analysis of Behavior* 79 (2): 267–86. https://doi.org/10.1901/jeab.2003.79-267.

Macho, Gabriele A., Cinzia Fornai, Christine Tardieu, Philip Hopley, Martin Haeusler, and Michel Toussaint. 2020. "The Partial Skeleton StW 431 from Sterkfontein – Is It Time to Rethink the Plio-Pleistocene Hominin Diversity in South Africa?" *Journal of Anthropological Sciences* 98: 73–88. https://doi.org/10.4436/JASS.98020.

MacKinnon, John. 1974. "The Behaviour and Ecology of Wild Orang-utans (*Pongo pygmaeus*)." *Animal Behaviour* 22 (1): 3–74. https://doi.org/10.1016/S0003-3472(74)80054-0.

Marivaux, Laurent, Francisco R. Negri, Pierre-Olivier Antoine, Narla S. Stutz, Fabien L. Condamine, Leonardo Kerber, François Pujos, Roberto Ventura Santos, André M.V. Alvim, and Annie S. Hsiou. 2023. "An Eosimiid Primate of South Asian Affinities in the Paleogene of Western Amazonia and the Origin of New World Monkeys." *Proceedings of the National Academy of Sciences* 120 (28): e2301338120. https://doi.org/10.1073/pnas.2301338120.

Marlowe, Frank W. 2006. "Central Place Provisioning: The Hadza as an Example." In *Feeding Ecology in Apes and Other Primates: Ecological, Physical and Behavioral Aspects*, edited by G. Hohmann, M.M. Robbins, and C. Boesch. Cambridge: Cambridge University Press.

McCoy, Lara. 2022. "The Good Trolls Fighting Climate Change." UPM, April 13. https://www.upm.com/articles/circular-economy/22/the-good-trolls-fighting-climate-change/.

McGrew, W.C. 2013. "Is Primate Tool Use Special? Chimpanzee and New Caledonian Crow Compared." *Philosophical Transactions of the Royal Society B: Biological Science* 368 (1630): 20120422. https://doi.org/10.1098/rstb.2012.0422.

McLennan, Matthew R. 2018. "'Tie One On': Nest Tying by Wild Chimpanzees at Bulindi – A Variant of a Universal Great Ape Behavior?" *Primates* 59 (3): 227–33. https://doi.org/10.1007/s10329-018-0658-7.

McNamara, Patrick, and Kelly Bulkeley. 2015. "Dreams as a Source of Supernatural Agent Concepts." *Frontiers in Psychology* 6: 283. https://psycnet.apa.org/doi/10.3389/fpsyg.2015.00283.

McNutt, Ellison J., Bernhard Zipfel, and Jeremy M. DeSilva. 2018. "The Evolution of the Human Foot." *Evolutionary Anthropology: Issues, News, and Reviews* 27 (5): 197–217. https://doi.org/10.1002/evan.21713.

McPherron, Shannon P., Zeresenay Alemseged, Curtis W. Marean, Jonathan G. Wynn, Denné Reed, Denis Geraads, René Bobe, and Hamdallah A. Béarat. 2010. "Evidence for Stone-Tool-Assisted Consumption of Animal Tissues before 3.39 Million Years Ago at Dikika, Ethiopia." *Nature* 466 (7308): 857–60. https://doi.org/10.1038/nature09248.

Mead, Margaret. 1928. *Coming of Age in Samoa*. New York: Morrow.

Milks, Annemieke. 2023. "Hominins Built with Wood 476,000 Years Ago." *Nature* 622 (7981): 34–6. https://doi.org/ 10.1038/d41586-023-02858-1.

Milks, Annemieke, Jens Lehmann, Utz Böhner, Dirk Leder, Tim Koddenberg, Michael Sietz, Matthias Vogel, and Thomas Terberger. 2022. "Wood Technology: A Glossary and Code for Analysis of Archaeological Wood from Stone Tool Cultures." *OSF Preprints*. https://doi.org/10.31219/osf.io/x8m4j.

Mills, Jon. 2020. "Toward a Theory of Myth." *Journal for the Theory of Social Behaviour* 50 (4): 410–24. https://doi.org/10.1111/jtsb.12249.

Mithen, Steven. 2024. *The Language Puzzle*. London: Profile Books.

Morwood, M.J., and W.L. Jungers. 2009. "Conclusions: Implications of the Liang Bua Excavations for Hominin Evolution and Biogeography." *Journal of Human Evolution* 57 (5): 640–8. https://doi.org/10.1016/j.jhevol.2009.08.003.

Morwood, M.J., R.P. Soejono, R.G. Roberts, T. Sutikna, C.S. Turney, K.E. Westaway, W.J. Rink, J.X. Zhao, G.D. van den Bergh, R.A. Due, D.R. Hobbs, M.W. Moore, M.I. Bird, and L.K. Fifield. 2004. "Archaeology and Age of a New Hominin from Flores in Eastern Indonesia." *Nature* 431 (7012): 1087–91. https://doi.org/10.1038/nature02956.

Murdock, George P., and Douglas R. White. 1969. "Standard Cross-Cultural Sample." *Ethnology* 8 (4): 329–69. https://doi.org/10.2307/3772907.

Nakamura, Michio, and Noriko Itoh. 2008. "Hunting with Tools by Mahale Chimpanzees." *Pan Africa News* 15 (1): 3–6. https://doi.org/10.5134/143489.

Napier, J.R. 1967. "The Antiquity of Human Walking." *Scientific American* 216 (4): 56–67. https://doi.org/10.1038/scientificamerican0467-56.

Newman, J.D. 2004. "Motherese by Any Other Name: Mother-Infant Communication in Non-hominin Mammals." *Behavioral and Brain Science* 27 (4): 519–20. https://doi.org/10.1017/S0140525X04400119.

Nishimura, Takeshi. 2018. "The Descended Larynx and the Descending Larynx." *Anthropological Science* 126 (1): 180301. https://doi.org/10.1537/ase.180301.

Nowack, Julia, A. Daniella Rojas, Gerhard Körtner, and Fritz Geiser. 2015. "Snoozing through the Storm: Torpor Use during a Natural Disaster." *Scientific Reports* 5 (1): 1–6. https://doi.org/10.1038/srep11243.

Nunn, C.L., and D.R. Samson. 2018. "Sleep in a Comparative Context: Investigating How Human Sleep Differs from Sleep in Other Primates." *American Journal of Physical Anthropology* 166 (3): 601–12. https://doi.org/10.1002/ajpa.23427.

Oller, D.K., E.H. Buder, H.L. Ramsdell, A.S. Warlaumont, L. Chorna, and R. Bakeman. 2013. "Functional Flexibility of Infant Vocalization and the Emergence of Language." *Proceedings of the National Academy of Sciences* 110 (16): 6318–23. https://doi.org/10.1073/pnas.1300337110.

Oller, D.K., M. Caskey, H. Yoo, E.R. Bene, Y. Jhang, C.C. Lee, D.D. Bowman, H.L. Long, E.H. Buder, and B. Vohr. 2019a. "Preterm and Full Term Infant Vocalization and the Origin of Language." *Scientific Reports* 9 (1): 14734. https://doi.org/10.1038/s41598-019-51352-0.

Oller, D.K., U. Griebel, S.N. Iyer, Y. Jhang, A.S. Warlaumont, R. Dale, and J. Call. 2019b. "Language Origins Viewed in Spontaneous and Interactive Vocal Rates of Human and Bonobo Infants." *Frontiers in Psychology* 10: 729. https://doi.org/10.3389/fpsyg.2019.00729.

Osvath, M., and G. Martin-Ordas. 2014. "The Future of Future-Oriented Cognition in Non-humans: Theory and the Empirical Case of the Great Apes."

Philosophical Transactions of the Royal Society B: Biological Science 369 (1655). https://doi.org/10.1098/rstb.2013.0486.

Ota, Mitsuhiko, Nicola Davies-Jenkins, and Barbora Skarabela. 2018. "Why Choo-Choo Is Better than Train: The Role of Register-Specific Words in Early Vocabulary Growth." *Cognitive Science* 42 (6): 1974–99. https://doi.org/10.1111/cogs.12628.

Ovando-Tellez, M., Y.N. Kenett, M. Benedek, M. Bernard, J. Belo, B. Beranger, T. Bieth, and E. Volle. 2022. "Brain Connectivity-Based Prediction of Real-Life Creativity Is Mediated by Semantic Memory Structure." *Science Advances* 8 (5): eabl4294. https://doi.org/10.1126/sciadv.abl4294.

Panjwani, Usha, Lalan Thakur, Jag Pervesh Anand, and Pratul Kumar Banerjee. 2007. "Sleep Architecture at 3500 Meters in a Sample of Indians." *Sleep and Biological Rhythms* 5 (3): 159–65. https://doi.org/10.1111/j.1479-8425.2007.00273.x.

Parker, Sue Taylor, and Kathleen R. Gibson. 1977. "Object Manipulation, Tool Use and Sensorimotor Intelligence as Feeding Adaptations in Cebus Monkeys and Great Apes." *Journal of Human Evolution* 6 (7): 623–41. https://doi.org/10.1016/S0047-2484(77)80135-8.

Pascual-Garrido, Alejandra, and Katarina Almeida-Warren. 2021. "Archaeology of the Perishable: Ecological Constraints and Cultural Variants in Chimpanzee Termite Fishing." *Current Anthropology* 62 (3): 333–62. https://doi.org/10.1086/713766.

Patel, Aakash K., Vamsi Reddy, and John F. Araujo. 2024. "Physiology, Sleep Stages." *StatPearls [Internet]*. https://www.ncbi.nlm.nih.gov/books/NBK526132/.

Perogamvros, Lampros, Anna Castelnovo, David Samson, and Thien Thanh Dang-Vu. 2020. "Failure of Fear Extinction in Insomnia: An Evolutionary Perspective." *Sleep Medicine Reviews* 51: 101277. https://doi.org/10.1016/j.smrv.2020.101277.

Petraglia, Michael, Emmanuel K. Ndiema, María Martinón-Torres, and Nicole Boivin. 2023. "Major New Research Claims Smaller-Brained *Homo naledi* Made Rock Art and Buried the Dead. But the Evidence Is Lacking." *The Conversation* (blog), June 5. https://theconversation.com/major-new-research-claims-smaller-brained-homo-naledi-made-rock-art-and-buried-the-dead-but-the-evidence-is-lacking-207000.

Pigati, Jeffrey S., Kathleen B. Springer, Jeffrey S. Honke, David Wahl, Marie R. Champagne, Susan R.H. Zimmerman, Harrison J. Gray, Vincent L. Santucci, Daniel Odess, and David Bustos. 2023. "Independent Age Estimates Resolve the Controversy of Ancient Human Footprints at White Sands." *Science* 382 (6666): 73–5. https://doi.org/10.1126/science.adh5007.

Plooij, F.X. 1984. *The Behavioral Development of Free-Living Chimpanzee Babies and Infants*. Norwood, NJ: Ablex.

Plummer, Thomas W., James S. Oliver, Emma M. Finestone, Peter W. Ditchfield, Laura C. Bishop, Scott A. Blumenthal, Cristina Lemorini, Isabella Caricola, Shara E. Bailey, and Andy I.R. Herries. 2023. "Expanded Geographic Distribution and Dietary Strategies of the Earliest Oldowan Hominins and Paranthropus." *Science* 379 (6632): 561–6. https://doi.org/10.1126/science.abo7452.

Pouw, Wim, and Susanne Fuchs. 2021. "Origins of Vocal-Entangled Gesture." *PsyArXiv*, November 19. https://doi.org/10.31234/osf.io/egnar.

Povinelli, Daniel. 2000. *Folk Physics for Apes: The Chimpanzee's Theory of How the World Works*. New York: Oxford University Press.

Povinelli, Daniel, and Nicholas G. Ballew. 2012. *World without Weight: Perspectives on an Alien Mind*. New York: Oxford University Press.

Povinelli, Daniel J., and John G.H. Cant. 1995. "Arboreal Clambering and the Evolution of Self-Conception." *The Quarterly Review of Biology* 70 (4): 393–421. https://doi.org/10.1086/419170.

Praetorius, Summer K., Jay R. Alder, Alan Condron, Alan C. Mix, Maureen H. Walczak, Beth E. Caissie, and Jon M. Erlandson. 2023. "Ice and Ocean Constraints on Early Human Migrations into North America along the Pacific Coast." *Proceedings of the National Academy of Sciences* 120 (7): e2208738120. https://doi.org/10.1073/pnas.2208738120.

Prasetyo, Didik, Marc Ancrenaz, Helen C. Morrogh-Bernard, S.S. Utami Atmoko, Serge A. Wich, and Carel P. van Schaik. 2009. "Nest Building in Orangutans." In *Orangutans: Geographical Variation in Behavioral Ecology*, edited by Serge A. Wich, S. Suci Utami Atmoko, Tatang Mitra Setia, and Carel P. van Shaik, 269–77. New York: Oxford University Press.

Prieur, J., and S. Pika. 2020. "Gorillas' (*Gorilla g. gorilla*) Knowledge of Conspecifics' Affordances: Intraspecific Social Tool Use for Food Acquisition." *Primates* 61 (4): 583–91. https://doi.org/10.1007/s10329-020-00805-6.

Pruetz, Jill D., and Paco Bertolani. 2007. "Savanna Chimpanzees, *Pan troglodytes verus*, Hunt with Tools." *Current Biology* 17 (5): 412–17. https://doi.org/10.1016/j.cub.2006.12.042.

Pruetz, Jill D., Paco Bertolani, K. Boyer Ontl, Stacy Lindshield, Mack Shelley, and Erin G. Wessling. 2015. "New Evidence on the Tool-Assisted Hunting Exhibited by Chimpanzees (*Pan troglodytes verus*) in a Savannah Habitat at Fongoli, Sénégal." *Royal Society Open Science* 2 (4): 140507. https://doi.org/10.1098/rsos.140507.

Raven, Han. 2019. "Notes on Molluscs from NW Borneo – Dispersal of Molluscs through Nipa Rafts." *The Festivus* 51 (1): 3–10. https://doi.org/10.54173/F511003.

Reinhardt, Kathleen D. 2020. "Wild Primate Sleep: Understanding Sleep in an Ecological Context." *Current Opinion in Physiology* 15: 238–44. https://doi.org/10.1016/j.cophys.2020.04.002.

Renn, Jürgen. 2019. "Learning from Kushim about the Origins of Writing and Farming." In *Culture and Cognition: Essays in Honor of Peter Damerow*, edited by Klaus Thoden, 11–27. Edition Open Access. Retrieved from https://eoa.hypotheses.org/116.

Revonsuo, Antti. 2000. "The Reinterpretation of Dreams: An Evolutionary Hypothesis of the Function of Dreaming." *Behavioral and Brain Sciences* 23 (6): 877–901.

Riley, James Whitcomb. 1885, November 15. "The Elf Child," title of poem later changed to "Little Orphant Annie." *Indianapolis Journal*.

Rizal, Yan, Kira E. Westaway, Yahdi Zaim, Gerrit D. van den Bergh, E. Arthur Bettis III, Michael J. Morwood, O. Frank Huffman, Rainer Grün, Renaud Joannes-Boyau, and Richard M. Bailey. 2020. "Last Appearance of *Homo erectus* at Ngandong, Java, 117,000–108,000 Years Ago." *Nature* 577 (7790): 381–5. https://doi.org/10.1038/s41586-019-1863-2.

Roberts, David L., Ivan Jarić, Stephen J. Lycett, Dylan Flicker, and Alastair Key. 2023. "*Homo floresiensis* and *Homo luzonensis* Are Not Temporally Exceptional Relative to *Homo erectus*." *Journal of Quaternary Science* 38 (4): 463–70. https://doi.org/10.1002/jqs.3498.

Róheim, Geza. 1933. "Women and Their Life in Central Australia." *The Journal of the Royal Anthropological Institute of Great Britain and Ireland* 63: 207–65. https://doi.org/10.2307/2843917.

Rosfort, R. 2013. "Folk Physics." In *Encyclopedia of Sciences and Religions*, edited by A.L.C. Runehov and L. Oviedo. Dordrecht: Springer.

Ross, C. 2001. "Park or Ride? Evolution of Infant Carrying in Primates." *International Journal of Primatology* 22: 749–71. https://doi.org/10.1023/A:1012065332758.

Roubik, David W. 2005. "Honeybees in Borneo." In *Pollination Ecology and the Rain Forest*, edited by David W. Roubik, Shoko Sakai, and Abang A. Hamid Karim, 89–103. New York: Springer.

Rousseau, Pierre V., Florence Matton, Renaud Lecuyer, and Willy Lahaye. 2017. "The Moro Reaction: More than a Reflex, a Ritualized Behavior of Nonverbal Communication." *Infant Behavior and Development* 46: 169–77. https://doi.org/10.1016/j.infbeh.2017.01.004.

Russon, Anne E., Dwi Putri Handayani, Purwo Kuncoro, and Agnes Ferisa. 2007. "Orangutan Leaf-Carrying for Nest-Building: Toward Unraveling Cultural

Processes." *Animal Cognition* 10 (2): 189–202. https://doi.org/10.1007/s10071-006-0058-z.

Ruxton, Graeme D., and David M. Wilkinson. 2012. "Population Trajectories for Accidental versus Planned Colonisation of Islands." *Journal of Human Evolution* 63 (3): 507–11. https://doi.org/10.1016/j.jhevol.2012.05.013.

Sammler, Daniela, Marie-Hélène Grosbras, Alfred Anwander, Patricia E.G. Bestelmeyer, and Pascal Belin. 2015. "Dorsal and Ventral Pathways for Prosody." *Current Biology* 25 (23): 3079–85. https://doi.org/10.1016/j.cub.2015.10.009.

Samson, David R. 2021. "The Human Sleep Paradox: The Unexpected Sleeping Habits of I." *Annual Review of Anthropology* 50: 259–74. https://doi.org/10.1146/annurev-anthro-010220-075523.

Samson, David R., Alyssa N. Crittenden, Ibrahim A. Mabulla, and Audax Z.P. Mabulla. 2017. "The Evolution of Human Sleep: Technological and Cultural Innovation Associated with Sleep-Wake Regulation among Hadza Hunter-Gatherers." *Journal of Human Evolution* 113: 91–102. https://doi.org/10.1016/j.jhevol.2017.08.005.

Samson, David R., and Kevin D. Hunt. 2012. "A Thermodynamic Comparison of Arboreal and Terrestrial Sleeping Sites for Dry-Habitat Chimpanzees (*Pan troglodytes schweinfurthii*) at the Toro-Semliki Wildlife Reserve, Uganda." *American Journal of Primatology* 74 (9): 811–18. https://doi.org/10.1002/ajp.22031.

Samson, David R., and Charles L. Nunn. 2015. "Sleep Intensity and the Evolution of Human Cognition." *Evolutionary Anthropology: Issues, News, and Reviews* 24 (6): 225–37. https://doi.org/10.1002/evan.21464.

Samuni, Liran, Franziska Wegdell, and Martin Surbeck. 2020. "Behavioural Diversity of Bonobo Prey Preference as a Potential Cultural Trait." *Elife* 9: e59191. https://doi.org/10.7554/eLife.59191.

Sanz, Crickette M., and David B. Morgan. 2009. "Flexible and Persistent Tool-Using Strategies in Honey-Gathering by Wild Chimpanzees." *International Journal of Primatology* 30 (3): 411–27. https://doi.org/10.1007/s10764-009-9350-5.

Savage-Rumbaugh, E. Sue, and William M. Fields. 2011. "The Evolution and the Rise of Human Language: Carry the Baby." In *Homo symbolicus: The Dawn of Language, Imagination and Spirituality*, edited by F. d'Errico and C.S. Henshilwood, 13–48. Amsterdam and Philadelphia: John Benjamins Publishing.

Scerri, Eleanor M.L., Patrick Roberts, S. Yoshi Maezumi, and Yadvinder Malhi. 2022. "Tropical Forests in the Deep Human Past." *Philosophical Transactions of the Royal Society B: Biological Sciences* 377 (1849): 20200500.

Schoch, Werner H., Gerlinde Bigga, Utz Böhner, Pascale Richter, and Thomas Terberger. 2015. "New Insights on the Wooden Weapons from the Paleolithic

Site of Schöningen." *Journal of Human Evolution* 89: 214–25. https://doi.org/10.1016/j.jhevol.2015.08.004.

Schredl, Michael. 2019. "Book Review: 'The Dreams behind the Music: Learn Creative Dreaming as 100+ Top Artists Reveal Their Breakthrough Inspirations' by Craig Sim Webb." *International Journal of Dream Research* 12 (2): 95. https://doi.org/10.11588/ijodr.2019.2.65363.

Seed, A., E. Seddon, B. Greene, and J. Call. 2012. "Chimpanzee 'Folk Physics': Bringing Failures into Focus." *Philosophical Transactions of the Royal Society B: Biological Sciences* 367 (1603): 2743–52. https://doi.org/10.1098/rstb.2012.0222.

Shultz, Susanne, Ronald Noë, W. Scott McGraw, and R.I.M. Dunbar. 2004. "A Community-Level Evaluation of the Impact of Prey Behavioural and Ecological Characteristics on Predator Diet Composition." *Proceedings of the Royal Society of London. Series B: Biological Sciences* 271 (1540): 725–32. https://doi.org/10.1098/rspb.2003.2626.

Shumaker, Robert W., Kristina R. Walkup, and Benjamin B. Beck. 2011. *Animal Tool Behavior: The Use and Manufacture of Tools by Animals*. Baltimore: Johns Hopkins University Press.

Simor, Péter, Gwen van der Wijk, Lino Nobili, and Philippe Peigneux. 2020. "The Microstructure of REM Sleep: Why Phasic and Tonic?" *Sleep Medicine Reviews* 52: 101305. https://doi.org/10.1016/j.smrv.2020.101305.

Sivakumar, Kuppusamy. 2010. "Impact of the Tsunami (December, 2004) on the Long Tailed Macaque of Nicobar Islands, India." *Hystrix, the Italian Journal of Mammalogy* 21 (1). https://doi.org/10.4404/hystrix-21.1-4484.

Spielman, A.J., and P.B. Glovinsky. 1991. "The Varied Nature of Insomnia." In *Case Studies in Insomnia*, edited by P.J. Hauri, 1–18. New York and London: Plenum Medical Book Co.

Stewart, Fiona A., Alexander K. Piel, Jurgi C. Azkarate, and Jill D. Pruetz. 2018. "Savanna Chimpanzees Adjust Sleeping Nest Architecture in Response to Local Weather Conditions." *American Journal of Physical Anthropology* 166 (3): 549–62. https://doi.org/10.1002/ajpa.23461.

Stewart, Fiona A., Alexander K. Piel, and W.C. McGrew. 2011. "Living Archaeology: Artefacts of Specific Nest Site Fidelity in Wild Chimpanzees." *Journal of Human Evolution* 61 (4): 388–95. https://doi.org/10.1016/j.jhevol.2011.05.005.

Stibbard-Hawkes, D.N.E. 2024. "Reconsidering the Link between Past Material Culture and Cognition in Light of Contemporary Hunter-Gatherer Material Use." *Behavioral and Brain Sciences* (March): 1–53. https://doi.org/10.1017/s0140525x24000062.

Suntsova, Maria V., and Anton A. Buzdin. 2020. "Differences between Human and Chimpanzee Genomes and Their Implications in Gene Expression, Protein Functions and Biochemical Properties of the Two Species." *BMC Genomics* 21 (7): 1–12. https://doi.org/10.1186/s12864-020-06962-8.

Surbeck, Martin, and Gottfried Hohmann. 2008. "Primate Hunting by Bonobos at LuiKotale, Salonga National Park." *Current Biology* 18 (19): R906–R907. https://doi.org/10.1016.j.cub.2008.08.040.

Sutikna, Thomas, Matthew W. Tocheri, Michael J. Morwood, E. Wahyu Saptomo, Rokus Due Awe, Sri Wasisto, Kira E. Westaway, Maxime Aubert, Bo Li, and Jian-xin Zhao. 2016. "Revised Stratigraphy and Chronology for *Homo floresiensis* at Liang Bua in Indonesia." *Nature* 532 (7599): 366–9. https://doi.org/10.1038/nature17179.

Tagg, Nikki, Jacob Willie, Charles-Albert Petre, and Olivia Haggis. 2013. "Ground Night Nesting in Chimpanzees: New Insights from Central Chimpanzees (*Pan troglodytes troglodytes*) in South-East Cameroon." *Folia Primatologica* 84 (6): 362–83. https://doi.org/10.1159/000353172.

Takemoto, Hiroyuki. 2004. "Seasonal Change in Terrestriality of Chimpanzees in Relation to Microclimate in the Tropical Forest." *American Journal of Physical Anthropology: The Official Publication of the American Association of Physical Anthropologists* 124 (1): 81–92. https://doi.org/10.1002/ajpa.10342.

Tanner, Nancy, and Adrienne Zihlman. 1976. "Women in Evolution. Part I: Innovation and Selection in Human Origins." *Signs: Journal of Women in Culture and Society* 1 (3, Part 1): 585–608. https://doi.org/10.1086/493245.

Teulier, C., D.K. Lee, and B.D. Ulrich. 2015. "Early Gait Development in Human Infants: Plasticity and Clinical Applications." *Developmental Psychobiology* 57 (4): 447–58. https://doi.org/10.1002/dev.21291.

Thach, Bradley T., and Anna Lijowska. 1996. "Arousals in Infants." *Sleep* 19 (suppl. 10): S271–S273. https://doi.org/10.1093/sleep/19.suppl_10.S271.

Timetree.org. Accessed November 12, 2023.

Tocheri, Matthew W. 2019. "Previously Unknown Human Species Found in Asia Raises Questions about Early Hominin Dispersals from Africa." *Nature* 568 (7751): 176–8. https://doi.org/10.1038/d41586-019-01019-7.

van Casteren, A., W.I. Sellers, S.K. Thorpe, S. Coward, R.H. Crompton, J.P. Myatt, and A.R. Ennos. 2012. "Nest-Building Orangutans Demonstrate Engineering Know-How to Produce Safe, Comfortable Beds." *Proceedings of the National Academy Sciences* 109 (18): 6873–7. https://doi.org/10.1073/pnas.1200902109.

Vančatová, Marina, and Václav Vančata. 2021. "The Simple Objects Place in Enclosure for Gorillas Initiate Rare or New Behavioural Patterns – Implications for the Origin of Hominine Tool Behaviour." *Anthropologie* 59 (1): 45–54. https://doi.org/10.26720/anthro.20.01.17.1.

Van den Bergh, Gerrit D., Yousuke Kaifu, Iwan Kurniawan, Reiko T. Kono, Adam Brumm, Erick Setiyabudi, Fachroel Aziz, and Michael J. Morwood. 2016. "*Homo floresiensis*-Like Fossils from the Early Middle Pleistocene of Flores." *Nature* 534 (7606): 245. https://doi.org/10.1038/nature17999.

Vonk, Jennifer. 2020. "Twenty Years after Folk Physics for Apes: Researchers' Understanding of How Nonhumans Understand the World." *Animal Behavior and Cognition* 7 (3): 264–9. https://doi.org/10.26451/abc.07.03.01.2020.

Vonk, Jennifer, and Michael J. Beran, eds. 2020. "The Twenty-Year Anniversary of Daniel Povinelli's (2000) Folk Physics for Apes: The Chimpanzee's Theory of How the World Works." Special Issue, *Animal Behavior and Cognition* 7 (3): 264–9.

Wadley, L., I. Esteban, P. de la Peña, M. Wojcieszak, D. Stratford, S. Lennox, F. d'Errico, D.E. Rosso, F. Orange, L. Backwell, and C. Sievers. 2020. "Fire and Grass-Bedding Construction 200 Thousand Years Ago at Border Cave, South Africa." *Science* 369 (6505): 863–6. https://doi.org/10.1126/science.abc7239.

Wagner, U., S. Gais, H. Haider, R. Verleger, and J. Born. 2004. "Sleep Inspires Insight." *Nature* 427 (6972): 352–5. https://doi.org/10.1038/nature02223.

Wakefield, Monica L., Alexana J. Hickmott, Colin M. Brand, Ian Y. Takaoka, Lindsey M. Meador, Michel T. Waller, and Frances J. White. 2019. "New Observations of Meat Eating and Sharing in Wild Bonobos (*Pan paniscus*) at Iyema, Lomako Forest Reserve, Democratic Republic of the Congo." *Folia Primatologica* 90 (3): 179–89. https://doi.org/10.1159/000496026.

Wall-Scheffler, C.M., and M.J. Myers. 2017. "The Biomechanical and Energetic Advantages of a Mediolaterally Wide Pelvis in Women." *The Anatomical Record* 300 (4): 764–75. https://doi.org/10.1002/ar.23553.

Watson, B.O., and G. Buzsáki. 2015. "Sleep, Memory and Brain Rhythms." *Daedalus* 144 (1): 67–82. https://doi.org/10.1162/DAED_a_00318.

Wermke, Kathleen, Michael P. Robb, and Philip J. Schluter. 2021. "Melody Complexity of Infants' Cry and Non-cry Vocalisations Increases across the First Six Months." *Scientific Reports* 11 (1): 1–11. https://doi.org/10.1038/s41598-021-83564-8.

Westaway, Michael C. 2019. "The First Hominin Fleet." *Nature Ecology & Evolution* 3 (7): 999–1000. https://doi.org/10.1038/s41559-019-0928-9.

Westaway, Michael Carrington, Arthur C. Durband, Colin P. Groves, and Mark Collard. 2015. "Mandibular Evidence Supports *Homo floresiensis* as a Distinct Species." *Proceedings of the National Academy of Sciences* 112 (7): E604–E605. https://doi.org/10.1073/pnas.1418997112.

Whiting, John W.M. 1981. "Environmental Constraints on Infant Care Practices." In *Handbook of Cross-Cultural Human Development*, edited by Ruth H. Munroe, Robert L. Munroe, and Beatrice Blyth Whiting, 155–79. New York and London: Garland STPM Press.

Whyte, Martin King. 1978. "Cross-Cultural Codes Dealing with the Relative Status of Women." *Ethnology* 17 (2): 211–37. https://doi.org/10.2307/3773145.

Wood, Brian M., and Ian C. Gilby. 2017. "From *Pan* to Man the Hunter: Hunting and Meat Sharing by Chimpanzees, Humans, and Our Common Ancestor." In *Chimpanzees and Human Evolution*, edited by Martin N. Muller, Richard W. Wrangham, and David R. Pilbeam, 339–82. Cambridge, MA: Harvard University Press.

Worthman, Carol M., and Melissa K. Melby. 2002. "Toward a Comparative Developmental Ecology of Human Sleep." In *Adolescent Sleep Patterns: Biological, Social, and Psychological Influences*, edited by M.A. Carskadon, 69–117. Cambridge: Cambridge University Press. https://psycnet.apa.org/doi/10.1017/CBO9780511499999.009.

Yetish, Gandhi, and Ronald McGregor. 2019. "Hunter-Gatherer Sleep and Novel Human Sleep Adaptations." In *Handbook of Behavioral Neuroscience: Handbook of Sleep Research*, edited by Hans C. Dringenberg, 317–31. Cambridge, MA: Elsevier.

Young, N.M., T.D. Capellini, N.T. Roach, and Z. Alemseged. 2015. "Fossil Hominin Shoulders Support an African Ape-Like Last Common Ancestor of Humans and Chimpanzees." *Proceedings of the National Academy of Sciences* 112 (38): 11829–34. https://doi.org/10.1073/pnas.1511220112.

Zhang, Lyubing, Eric I. Ameca, Guy Cowlishaw, Nathalie Pettorelli, Wendy Foden, and Georgina M. Mace. 2019. "Global Assessment of Primate Vulnerability to Extreme Climatic Events." *Nature Climate Change* 9 (7): 554–61. https://doi.org/10.1038/s41558-019-0508-7.

Zihlman, Adrienne L. 1981. "Women as Shapers of the Human Adaptation." In *Woman the Gatherer*, edited by F. Dahlbert, 75–120. New Haven, CT: Yale University Press.

Zihlman, Adrienne L. 1985. "Gathering Stories for Hunting Human Nature." *Feminist Studies* 11: 365–77. https://www.jstor.org/stable/3177929.

Zihlman, Adrienne L., and Carla Simmons (illustrator). 2000. *The Human Evolution Coloring Book*. New York: Coloring Concepts.

Zihlman, Adrienne L., and Carol E. Underwood. 2019. *Ape Anatomy and Evolution*. Self-published.

Index

Page numbers in italics refer to figures.

!Kung San, 57, 77, 204n16

Acheulean tools, 93–6, *94*, 110, 143, 181. *See also* stone tools
Adolph, Karen E., 32–3
Africa: forest restructuring in, 42; origin of hominins, 9. *See also* Late Miocene Cooling
Anderson, Helen, 87–9, 147, 172–6, *173*, 205n6
arboreal sleeping nests: baskets in trees, 6, 15, *16*, 25, 48, 50, 127, 136, 172, 174; of early hominins, 5, 6, 10, 13, 15, 52, 70, 142; functions, 14, 17, 137; of great apes, 5, 10, 13–14, 16, 51, 56; how and why great apes build them, 13–17, 193n23; of *Homo floresiensis*, 128, 131, 135, 182; independently invented by great apes as stationary tools, 13; nesting materials, 46–8; as possible life rafts, 132–6, 211n22. *See also* baskets; nest building
arboreality: in orangutans, 13–17, 24, 27, 30, 48, 51; in chimpanzees and gorillas, 14, 51
archaeological sites, 32, 59, 87, 89, 93, 95–6, 96, 103, 105, 147, 166, 180, 181
australopithecines, 51–2, 63, 68–70, 74–6, 128, 131, 164–6
Australopithecus afarensis, 38–9, 53, 70, 75–6, 93
Australopithecus africanus, *52*, 197n1
Australopithecus prometheus, *37*, 36–40. *See also* Little Foot

baby cradles: compared to baby slings, 78–82, *81*; in colder climates, *80*;

baby cradles (*continued*)
    oldest in archaeological record, 87–8, *88*; made from vegetal matter, 91. *See also* botanical tools
baby slings: derived from arboreal sleeping nests, 6, 69, 71, 74–5, 127; facilitators of evolution, 109, 112–13, 125–6, 144, 168–71, 182; invented during Botanic Age, 6, 74–6, 87, 108, 110, 112–14, 136, 139, 142, 144, 203n12; as means to free maternal hands, 44, 47, 74, 171, 204n19; modern use of, *43*, 69, 76–82, 79–80, 85, 113; as precursors to gathering bags, 47, 82–3, 86; in warmer climates, 80. *See also* botanical tools
baskets: basket weaving as most ancient craft, 147; inferred in archaeological record from crosshatched designs, 175; on the ground, 48, 127; on the hips, 126–7; in the seas, 127, 132–6, *134*, 136, 138; widespread in humans, 174. *See also* arboreal sleeping nests; weaving
Beck, Benjamin, 11–12
beds: at Border Cave, South Africa, 59; bedrooms, 58–9; bedtimes in different cultures, 57, 59; bedtime stories, 151–2; and central place sleeping, 83; children's fears of what's under, 63–4, 202n57; of grass, 96, 174; human beds derived from tree nests, 50; made in morning by humans unlike apes, 10, 50; shared with infants, 78; terrestrial beds of botanical matter, 67. *See also* arboreal sleeping nests; nest building
Bertolani, Paco, 101, *102*
bipedalism: in apes and humans, 27–32; distinguishes hominins from apes, 5; driver for language evolution, 20, 122–7, 154–60, *158*; evolution of, 5, 13, 41, 70, 118–19, 126; loss of grasping toes, 68, 70–2, 96, 111, 126; means of freeing hands, 44, 47, 74, 171, 204n19; morphological features allowing for, 28, *29*, 30, 36–40. *See also* rhythm and bipedalism
bonobos. *See* chimpanzees
Border Cave, 32. *See also* archaeological sites
Botanic Age: defined, 6, 25, 40, *41*, 198n26; evolution of bipedalism in, 41, 44, 70, 118–19, 126; evolution of hominins during, 7, 41, *41*, 66–7, 112–14, 117, 119; "formative years" of hominins, 5; hominin global expansion during, 76, 80, *130*, 136, 213n50; language development in, 122–3; shift to terrestrial life during, 49, 103; tool use during, 58, 75, 81–2, 91, 108, 111, 139–40, 142, 148
botanic matter, reverence for: trees and botanic items in folklores and religion, 145, 214n5, 214–15n6; veneration by environmentalists and artists, 145–9, *146*, *147*, *149*. *See also* García, Louie
botanical tools: as contributing to human evolution, 6–7, 142–8; paper, 7, 139–41; textiles, 87–90; wood tools, 91–2. *See also* baby slings

chimpanzees: bonobos, *9*, 14, 98, 115, 157, 192n2, 193n11; closest human relative, 4–5, 9; cognitive skill development, 18–20; developmental milestones compared to humans, 33–5, *34*; folk (intuitive) physics in, 18–25, 194n37, 195n47, 195n48, 196–7n56; food, 107; gendered roles, 98–103, 108–9; ground nesting, 44–5, *45*, 160–2; hunting, 97–102, 206n31; infants carried by mothers, 72, *73*; nest building, 13–17, *16*, 25, 44–7, *45*, 160–3, 193n23; nesting materials, 46–8; savanna chimpanzees, 46, 100–3, 105–6, 109; sleep patterns, 51, 56, 60, 134, 201n28; social communities, 82, 156–8; tool use, 10–13, 19–25, 97–103, 193n7, 206n25; uneven walking rhythms of, 27, *29*, 30, *31*, 125, 157; weaving in one individual, 25, 194n29. *See also* great apes

Clacton spear, 105–6, *106*

Clarke, Ronald, 30, 36–7, *37*

cognitive leaps, 69, 93–4, 96, 110, 127, 143, 180

common ancestor of chimpanzees and humans, *9*, 20, 41, 193n18

Coolidge, Frederick, 93–4

Coss, Richard, 63–4, 202n57

Dambo, Thomas, 145–6, *146*

Darwin, Charles, 3, 111, 192n1

DeSilva, Jeremy, 38

Dikika baby. See *Australopithecus afarensis*

Dougherty, Patrick and Sam, 146–7, *147*

Edison, Thomas, 62

Ennos, Roland, 137, 198n26, 206n25, 212n47, 213n3

entrainment: ability to keep beat to rhythmic sounds, 124–5, 210n50; in humans, 154–6; lacking in chimpanzees and other primates, 156–7; in music and language, 158–9. *See also* rhythm and bipedalism; rhythm and language

environment: changes during Botanic Age, 41–3; as driver for terrestrial sleeping, 46–7, 67, 71; of early hominins, 165, 199n34, 199n37; environmentalists, 145–8; of savanna chimpanzees, 46, 100; and sleeping nests, 194n30; sounds from, 155. *See also* Late Miocene Cooling

evo-devo (evolutionary developmental biology), 31, *34*, 118–19, 193n24, 208n13

fear of heights as adaptive, 17, 39, 194n31

Filippi, Piera, 123

folk physics: also called intuitive or naïve physics, 6, 17–18, 20, 22, 74; in chimpanzees, 18–25, 194n37, 195n47, 196n56; in early hominins, 6, 50, 68, 74, 142; in humans, 5, 17–20, 22; and invention of baby slings, 136

foot: aligned big toes, 30, 35, 68, 70; big toe not in chimpanzee feet, 30, *31*; evolved first, 40, 71, 126; foot anatomy, 30, 40, 53; footprints, 139, 198n20; in fossils hominins, 36, 38, 49, 68, 70, 75, 131; four-footed, 5, 27; grasping in great apes, 30, 73, 75; Little Foot, 36–9, *37*, 52; migrating on foot, 138; mothers' footsteps, 125, 154–6, 158–9, *158*; plantar grasp in babies, 115; tapping to a beat, 124; role in walking, 28, 30, 32, 35, 71; weight-bearing in bipedal hominins, 30, 71, 128, 191n4

fossil record: of australopithecine, 36–40, 51, 68, 75–6, 164, 166, 168; of botanical tools, 76, 87, 88–9, 147–8, 172–6; evolution of bipedalism inferred from, 35, 39–40, 53, 68; of hominins, 4–5, 9–10, 26, 33, 36–41, 63, 70–1. *See also* australopithecines; hominins

García, Louie, 148, *149*, 186–90, *187*
gendered roles: in chimpanzees, 98–103, 108–9; in humans, 82–6, 105–9, 168–72. See also women as "the burden-bearing sex"
Gönnersdorf site, 87, *88*. See also archaeological sites
Goodall, Jane, 10, 15, 48, 73, 100, 193n7
gorillas, 5, *9*, 11, 14, 15, 27, 30, 42, 44, 51, 63, 72. See also great apes
grasping reflex: in *Australopithecus afarensis*, 75; in chimpanzees, 72–3, 115; in hominins, 74; in humans, 115–16, *116*, 208n18. See also infant clinging
great apes: anatomy, 13, 28, *29*, 30, *31*; closest primates to humans, 8, *9*; different personalities in three great apes, 11–12; evolutionary timeline, *9*, 8–17, 70, 192n2, 192n3, 192n4; grasping big toes of, 30, *31*, 68, 70–2, 75; grasping hands, 70–1, *72*; lack tails, 9; large bodies of, 8, 193n18; tool use in, 10–13, 19–25, 82, 97–103, 193n18, 206n25; upper body strength, 30, 70–1, 193n18. See also chimpanzees; gorillas; orangutans
Greenwald, Alexandra, 80–1
Gürbüz, Rebecca Biermann, 96, 177–82, *178*

Hadza, 53, 58, *58*, 107
Hobbit. See *Homo floresiensis*
hominins (early): anatomy of bipedalism, 68, 70–2, 76, 111, 126; during the Botanic Age, 4–13, 41, 118–19, 198n26; brain evolution of, 33, 111, 122–7, 144; characteristics of, 9–10; diverged from chimpanzee lineage, 4, *9*;

evolutionary timeline, *9*; feet of, 30, *31*, 71, 76, 126; folk physics in, 6; habitual bipedalism of, 5, 27–8, 35, 38–40, 53, 70, 157, 168; hips of, 13, 28, *29*, 30; hunting in, xiv, 97, 103, 105–8, 148, 181–2; motor development, 32–5, *34*; sleep, 15, 60, 63, 66–7, 94; teeth, 26–7; thigh bones, 28, 30; tool use, 10–13, 97, 103–8, 181
Homo erectus, 71–2, 75, 89, 93–5, 112–13, 128, 131–2, 162, 173
Homo floresiensis, 128–32, *129*, *130*, 182
Homo habilis, 128, 211n4
Homo luzonensis, *130*, 131–2, 135
Homo sapiens, 4, 17, 108, 111, 125, 127, 129, 138, 170
human evolution, *9*, 3–7, 8–13, 34, 59, 105, 195–6n48, 206n25
hunting: in chimpanzees, 97–102, *102*; in contemporary hunting and gathering communities, 56–7, 107–8, 169, 176. See also chimpanzees; hominins

infant clinging: in apes and monkeys, 5, 71, *72*, *73*, 115; evolutionary loss in human babies, 44, 115; loss related to invention of baby slings, 112–13; significance of loss for hominin evolution, 52, 68, 71–4, 117–19, 126, 207n1. See also motor reflexes; putting the baby down (PTBD) theory

Just So Stories, 151–2, *152*

Kappelman, John, 39
Kehoe, Alice, 89–90
Kipling, Rudyard, 151–3, *152*
Köhler, Wolfgang, 20–5, 193n7, 193n23, 194n29, 195n43, 195n45, 195n47, 195–6n48, 197n57

Koops, Kathelijne, *16*, 44–5, *45*, 160–3, *161*

language: evolution of, 110–12, 122, 167, 207n1, 209n39; invention of, 111–12, 126–7; learning, 120–6, 154–60, 193n24; motherese, 111, 119–22, 126, 154–9. *See also* putting the baby down (PTBD) theory; rhythm and language
Larsson, Matz, 125–6, 154–60, *155*
Late Miocene Cooling, 41–2, 47, 67, 103
LB1. See *Homo floresiensis*
Lindsay, David, 117–19, 122, 208n11
Little Foot, 36–9, *37*, 52, 70, 198n19, 202n55. See also *Australopithecus prometheus*
Lomekwi site, 93, 96, 180, 205n11. *See also* archaeological sites
Lucy, 70. See also *Australopithecus afarensis*

Madam Bee: chimpanzee mother of Bee-hinde, 73–4, 119. *See also* Goodall, Jane
Man the Hunter, Woman the Gatherer, 107–8
Marlowe, Frank, 83
Marshall, Glenn, 136, 182–6, *183*, *184*
Mead, Margaret, xiv–xv, 191n1
Metal Ages, 140
Miocene. *See* Late Miocene Cooling
Money, Percy John, *79*, *85*, 86
motherese, 111, 119–22, 126, 154–9. *See also* language
motor reflexes: in bonobos, 115; in chimpanzees, 72–3; in human babies, 33, 115–16, *116*; reflex vocalizations, 123; as vestiges from the past, 114–18, 208n18. *See also* Lindsay, David

Neolithic: beginning of, 137, 191n1; inventions of reading, writing, and paper, 140. *See also* Stone Age
nest building: baby slings derived from, 69, 71, 73–6, 142; in chimpanzees, *16*, 212n40; collecting nest building materials bipedally, 44, 47–8, 67; in early hominins, 15, 70, 142; in great apes but not monkeys, 5, 13, 174; ground (terrestrial) nests, 14–15, 44–9, *45*, 50–3, 60, 63, 66–8, 71, 113, 127, 160–3, 167; human bed making, 50; innate and learned components of, 15, 193n23, 193n24; nest materials, 6, 41, 44, 46–8, 51, 98, 174, 194n30; nest tying, 16, 194n29; in orangutans, 15–16, 212n40, 212n41; starting point for advanced cognition, 12–13; weaving nests, 6, 25, 142. *See also* arboreal sleeping nests; baskets

Oldowan tools, 93, 96, 206n22. *See also* stone tools
orangutans: evolutionary relationship to chimpanzees, gorillas, and humans 5, *9*, 135; leaf-carrying in, 48; most arboreal great ape, 14, 27, 48; nest building, 5, 13–15, 48, 134–5, 174, 178, 212n40; shelter in trees during storms, 134–5, 212n40, n41; tool use in, 11–12, 178; quadrumanous, 27, 197n2. *See also* great apes
Ota, Mitsuhiko, 121–2

Paleolithic. *See* Stone Age
paper: invented in China, 140
paradoxical sleep: REM sleep, 55, 202n40; sleep paralysis, 55, 59, 202n47. *See also* sleep

Peninj site, 95–6, 206n22. *See also* archaeological sites
Plooij, Frans, 73–4
Povinelli, Daniel, 19–20
predation: on early hominins, 17, 51, 163–8; and fire, 60, 200n11; and ground sleeping, 46, 67–8, 83, 162; and nightmares, 63–4, 202–3n57; predators of early hominins, 51, *52*; vulnerable during REM sleep, 60. *See also* Shultz, Susanne
Pruetz, Jill, 101, *102*
putting the baby down (PTBD) theory, 111–14, 117, 119, 122. *See also* motherese; motor reflexes

rafts: colonization by animals on, 132–4, *134*, 137; colonization by hominins of Flores and Luzon, 131–2, 135–6; experiments with, 135–6, 182–6, *183, 184*; palms as rafts, 133, *134*; rafting to Australia and Flores, 182–6; sleeping nests as rafts, 127, 132, 134–5. *See also* baskets (in the seas); Marshall, Glenn
reading and writing: invented around 5,500 years ago, 140
reflexes. *See* motor reflexes
rhythm and bipedalism, 122–7, 154–60. *See also* bipedalism
rhythm and language, 120, 123–4. *See also* language
Richter, Joachim, *158*, 125–6

Samson, David, 59–60
Schöningen, 103, *104*, 105, 181. *See also* archaeological sites
seacraft, 136–9, *137*, 182–6
Shultz, Susanne, 51, 163–8, *164*
SK 54, 166. *See also* predation (on early hominins)
sleep: creativity during, 55, 61–2; dreams and nightmares, 55, 58, 61–6, *65*; evolution of, 53–6, 59–60, 63–6, 116–17, 163; functions of, 55; in great apes, 10, 14–15, 56, 201n28; group sleeping to avoid predators, 53; patterns in traditional societies, 56–9, *58*, 201n22; physiological stages in humans, 54–5, *54*, 200n18, 201n28; REM (rapid eye movement) sleep, 54–6, *54*, 59–62, 68, 202n40; sleep paralysis, 55, 59, 202n47. *See also* arboreal sleeping nests; paradoxical sleep
spears: chimpanzees use to hunt, 100–2, *102*, 106, 109; and digging sticks, 103; earliest made of wood, 95, 105, *106*, 143; thrusting versus throwing, 106; used by men more than women, 85–6, *85*, 106, 108. *See also* sticks
sticks: used during Botanic Age, 6; broad use by chimpanzees, 19, 21–2, 25, 98, 103; chimpanzee extractive foraging with, 100; chimpanzees hunt with, 100–2, *102*, 178; Swiss Army knives of chimpanzees, 21, 98; chimpanzees use as weapons, 168; digging sticks, 103; digging sticks used by women more than men, 105–6, 108; related to spears, 103; used by early hominins, 51, 95, 103, *104*, 108, 181; used by great apes, 25, 98. *See also* spears
Stone Age: after Botanic Age, xiii, 6–7, 25, *41*, 142; boundary with Botanic Age likely to change, 93; divisions of, 191n1; farming during, 137; metal ages, 140; Neolithic (New Stone Age), 137; stone tools made during, 3, 40, 94–8, *94*; wooden artifacts continued to be made during, 91–2, *91*, 95, 103–4, *104*, 143. *See also* Gürbüz, Rebecca Biermann; stone tools

stone tools: Acheulean tools, 93–6, *94*, 110, 143, 181; and human evolution, 3–4, 40, 75–6, 92–7, 108–9; oldest, 75, 143; Oldowan tools, 93, 96; presumption (biased) of superiority over botanical tools, 110, 126–7, 148; stone flakes, 93, 96, 180–1; use to refine wood, 95–7, 103, 105, 177–82, 206n22. *See also* Stone Age

Tanner, Nancy, 74, 82
Taung, *52*, 166. See also *Australopithecus africanus*
terrestrial life, adapting to, 20, 45–7, 164–8
terrestrial nests. *See* nest building
toes. *See* foot
tool use: aimed throwing in chimpanzees, 24, 98, 206n39; in animals, 10; botanic inventions to transport things, 91; botanic tools in early hominins, 13, 97, 108, 181; in chimpanzees compared to other apes, 10–12, 193n7; digging sticks of savanna chimpanzees, 103; for feeding, 10–11, 19–21, 24–5, 95–8, 101–3; in great apes, 10–11, 98, 168, 178–9. *See also* arboreal sleeping nests; stone tools
tree nests. *See* arboreal sleeping nests

Wall-Scheffler, Cara, 83–4, 168–72, *169*
weaving: in art, 146–8, *149*, 186–9, *187*; baby carriers, 76, 80, 87; "Basket Weaving 101," 6, 7, 148, 191n6; in a chimpanzee, 25, 194n29; of great ape nests, 5–6, 15–16, 25, 47, 67, 74–6, 80, 142; in hominins, 15, 25, 74, 76, 87–90, 146–8, 172–5. *See also* Anderson, Helen; baskets; García, Louie
Westaway, Michael, 129
Whiting, John Wesley Mayhew, 77–8, *80*
women as "the burden-bearing sex," 82–6, 168–72. *See also* gendered roles
Wonderwerk Cave, 89, 147. *See also* archaeological sites
wood artifacts: art, 145–7, *146*, *147*; canoes, 91, *91*, *137*; and cognition, 177–82; cradles, 79; notched logs, 105, 143; oldest in record (polished plank), 105, 143; shelters, *58*, 91, *91*, 139; ships, 138–9; spears, 95, 103–6, *104, 106*, 108, 143; wheels, 91, *92*. *See also* Ennos, Roland; Gürbüz, Rebecca Biermann
writing, 140, 213n57
Wynn, Thomas, 93–4

Zihlman, Adrienne, 74, 82, 197n5

www.ingramcontent.com/pod-product-compliance
Ingram Content Group UK Ltd.
Pitfield, Milton Keynes, MK11 3LW, UK
UKHW040026100325
455904UK00007B/9/J